C0-AVV-190

Eisenhower and the Missile Gap

A book in the series

CORNELL STUDIES IN SECURITY AFFAIRS

edited by Robert J. Art, Robert Jervis,
and Stephen M. Walt

A complete list of series titles appears at the end of the book.

Eisenhower and the Missile Gap

PETER J. ROMAN

Cornell University Press

ITHACA AND LONDON

First published 1995 by Cornell University Press.

Printed in the United States of America

Library of Congress Cataloging-in-Publication Data

Roman, Peter J.
Eisenhower and the missile gap / Peter J. Roman.
p. cm.—(Cornell studies in security affairs)
Includes bibliographical references and index.
ISBN 0-8014-2797-5 (cloth: alk. paper)
1. Nuclear weapons—Government policy—United States—History. 2. Executive power—United States—History. 3. United States—Politics and government—1953–1961. I. Title. II. Series.
UA23.R63 1995
358.1'7'0973—dc20 95-32731

The paper in this book meets the minimum requirements of the American National Standard for Information Sciences—Permanence of Paper for Printed Library Materials, ANSI Z39.48-1984.

Contents

Acknowledgments

I have been fortunate in researching and writing this book to receive encouragement and support from a number of institutions and individuals. The staffs of the Dwight D. Eisenhower Library, the John F. Kennedy Presidential Library, the National Archives, the National Security Archive, the Air Force Historical Research Center, the Harvard University Law School Library, the Library of Congress, and the Air Force Academy Library provided great help locating the necessary documents. David Haight of the Eisenhower Library and Duane Reed of the Air Force Academy were especially generous with their knowledge of their library's collections. The University of Wisconsin–Madison, the University of Colorado–Boulder, and the Kennedy Library supported my research with travel grants.

I thank my colleagues who shared their time and insights. The comments of Vaughn Altemus, Stephen Cimbala, and Edward Coffman on early drafts improved the book considerably. Allen Greb, David Goldfischer, Tari Renner, and Steve Borrelli helped me clarify certain concepts. Robert Art was instrumental in developing this book; his advice and commentary improved it immeasurably. David Tarr deserves special recognition for his support throughout the project. His generous counsel over the years, his encouragement, and his thorough commentary on previous drafts have been invaluable.

Finally, I would like to thank my friends and family for their support, particularly my brothers and their wives, the Buccis, the Dickies, Fredda Ruppenthal, and Rhonda Norman. My parents, Joyce and John Roman, instilled in me the desire to learn and taught me the importance of completing tasks. These values, and their constant support, were a source of strength as I worked on this book.

This book is dedicated to the memory of two people: my aunt, Lena Sommo, and my research assistant at the University of Alabama, Tom Barnes. Both passed away suddenly, before the book was completed. I hope each would have been proud of the final product.

P. J. R.

Abbreviations

AASM	Advanced air-to-surface missile
ABM	antiballistic missile
ACW	Ann C. Whitman
AFAL	Air Force Academy Library
AFHRC	Air Force Historical Research Center
AICBM	anti-intercontinental ballistic missile
ALBM	air-launched ballistic missile
ARPA	Advanced Research Projects Agency
ASM	air-to-surface missile
ASW	antisubmarine warfare
BMD	ballistic missile defense
BMEWS	Ballistic Missile Early Warning System
BNSP	Basic national security policy
BOB	Bureau of the Budget
CAMAL	Continuously Airborne Missile Launcher
CEP	circular error probable
CINCLANT	Commander-in-Chief, Atlantic
CINCSAC	Commander-in-Chief, Strategic Air Command
CIA-OSI	Central Intelligence Agency, Office of Scientific Intelligence
CJCS	Chairman, Joint Chiefs of Staff
CNO	Chief of Naval Operations
CSUSA	Chief of Staff, United States Army
CSUSAF	Chief of Staff, United States Air Force
DCI	Director, Central Intelligence
DDEL	Dwight D. Eisenhower Library
FOIA	Freedom of Information Act
FY	Fiscal Year
IOC	Initial operational capability
JCS	Joint Chiefs of Staff

JFKL	John F. Kennedy Library
MAD	mutual assured destruction
MAFB	Maxwell Air Force Base
MCP	Memorandum of Conference with the President
MMB	Modern Military Branch
MMP	Memorandum of meeting with the president
NATO	North Atlantic Treaty Organization
NESC	Net Evaluation Subcommittee
NIE	National Intelligence Estimate
NSA	National Security Archive
NSC	National Security Council
ODM-SAC	Office of Defense Mobilization, Scientific Advisory Committee
ONE	Office of National Estimates
OSANSA	Office of the Special Assistant for National Security Affairs
OSD	Office of the Secretary of Defense
OSS	Office of the Staff Secretary
OSST	Office of the Special Assistant to the President for Science and Technology
PSAC	President's Science Advisory Committee
RG	Record Group
SAC	Strategic Air Command
SACEUR	Supreme Allied Commander, Europe
SIOP	single integrated operational plan
SLBM	Submarine-launched ballistic missile
SOSUS	Sound surveillance system
SSBN	strategic nuclear ballistic missile submarine
USSC	United States Strategic Command
WHO	White House Office
WSEG	Weapons Systems Evaluation Group

Eisenhower and the Missile Gap

[1]

The Eisenhower Administration, Presidential Power, and Nuclear Policy

On October 4, 1957, the Soviet Union launched Sputnik I and placed the first man-made object into orbit around the Earth. As the 184-pound sphere beeped innocently in outer space, the United States struggled to make sense of this momentous occasion. Scientifically, Sputnik realized humankind's ageless dream of escaping the Earth's atmosphere. The event marked the Soviet Union's contribution to the International Geophysical Year—an effort in many countries, including the United States, to promote the scientific exploration of space. But Sputnik presented ominous military implications for the United States. Over the previous decade, America's lead over the Soviet Union in the development of new strategic and nuclear weapons had declined. Now it seemed the Soviets had overtaken the United States in developing a revolutionary technology that had obvious military applications. Both government officials, despite being privy to intelligence estimates detailing the Soviet missile program, and the American public were shocked by news of the Soviet Union's mastery of rocketry. Perhaps of greatest concern within the government was the size of the satellite, since any rocket large enough to lift Sputnik into orbit might also be able to lift a nuclear warhead. Soviet boasts of an intercontinental ballistic missile (ICBM) test only two months earlier now took on a grave new light. Development of such a weapon by America's Cold War enemy could undermine U.S. military superiority and threaten national security.

In the month that followed the launch of Sputnik I, President Dwight D. Eisenhower and administration officials tried to quell the uproar and reassure the public of the effectiveness of U.S. defense and missile policies. But two additional shocks followed in early November. First, the Soviet Union launched Sputnik II, which carried a live dog into space. While the United States still had not fired a successful space launch, the Soviets had fired two and had developed a rudimentary life-support system. The second shock came in the form of a report by the Security Resources Panel of the Office of Defense Mobilization-Science Advisory Committee (ODM-SAC). Chaired by H. Rowan Gaither, the panel had been organized by the National Security Council (NSC) to investigate the contributions of active and passive defense to deterring a nuclear war and ensuring the nation's survival should one occur. It concluded that active and passive defenses could provide "no significant protection"; "the protection of the United States and its population rests, therefore, primarily upon the deterrence provided by SAC [Strategic Air Command]."[1] Security would be difficult to achieve in the short term because of the Soviet Union's aggressive foreign policy and weapons advances. Consequently, the panel proposed two sets of recommendations for structuring U.S. defense programs: (1) "highest priority measures," which sought to reduce SAC vulnerability and increase striking capability, and (2) "less than highest priority measures," which addressed passive and active defenses. Cumulatively, the Gaither panel's recommendations required increasing defense expenditures by $44.22 billion over the next five years: $19.09 billion for highest priority measures and $25.13 billion for the other measures.[2] Increases in defense spending of this magnitude threatened President Eisenhower's policy of fiscal control and meant that total defense spending might exceed his $38 billion ceiling.

Only weeks after the Gaither report, the Central Intelligence Agency completed its first post-Sputnik assessment of Soviet missile programs. In its intelligence evaluation the CIA warned that the Soviet Union might be able to deploy hundreds of ICBMs within as little as two years. Such large deployments could shift the strategic balance in favor of the Soviet Union if the United States did not preserve its capability to retaliate by either deploying new missile forces or protecting existing forces. The "missile gap" now seemed a reality.

The missile gap period constitutes a critical stage in the evolution of U.S. strategic nuclear policy.[3] The nation's nuclear forces were now confronted with the prospect of vulnerability to a Soviet missile attack, although the severity of the threat remained unclear because of inconclusive intelligence. The Eisenhower administration reassessed the nation's strategic nuclear policies and developed new policies, weapons, and procedures to decrease the vulnerability of U.S. nuclear forces. Eisenhower's choices set the foundation for how the United States would cope with strategic vulnerability for the remainder of the Cold War. Although some policy changes remained incomplete when Eisenhower left office, his decisions during the missile gap period mark a key stage in the transition from strategic policies based on nuclear superiority to ones based on mutual vulnerability.

This book examines the Eisenhower administration's strategic nuclear policy-making processes and choices during the missile gap period. Most political scientists and historians ignore the strategic nuclear policies of the Eisenhower administration during this period and focus instead on those of the Kennedy administration and Secretary of Defense Robert McNamara. This is an unfortunate oversight. Even though relatively stable budgets projected—erroneously—an image of complacency, American strategic nuclear policies and programs underwent dramatic changes during the last three years of the Eisenhower administration. By investigating key areas of American strategic nuclear policy, this book shows the importance of the Eisenhower administration in setting the nation's strategic nuclear policy in an era of vulnerability.

In addition to analyzing the content of Eisenhower's strategic nuclear policies, this book examines the policy processes behind these decisions. Most, although not all, of this examination of strategic nuclear policy concentrates on the highest levels of government—the president, the NSC, the secretaries of state and defense, and the Joint Chiefs of Staff (JCS). Consequently, this book provides an evaluation that informs both the general literature on presidential power in foreign policy making and perspectives on the Eisenhower presidency.

Investigating the Eisenhower administration's decision making on strategic nuclear policy during the missile gap period is possible only because of the declassification of thousands of government documents over the past fifteen years. This book is based primarily on the document collections of the Dwight D. Eisenhower Library, the John F. Kennedy Presidential Library, and the National Archives, as well as the various armed services archive collections. The availability of declassified records has enabled me to recreate the government's internal policy process and better understand how the administration began adjusting to an era of vulnerability. In some cases, such as budgets and force planning, the results of debates became known immediately. In others, such as intelligence analysis and nuclear strategy, the administration's policies were either unknown or only vaguely known even for decades after the end of the missile gap period (1961). The policy processes and internal debates in all these areas have been virtually unknown until now.

Presidential Power, Foreign Policy Making, and Eisenhower

Before we assess President Eisenhower's decisions with regard to strategic nuclear policy during the missile gap period, we should acknowledge two overlapping areas of scholarship: presidential power in foreign policy decision making and the Eisenhower presidency. A comprehensive review of the literature of each would be a difficult undertaking given their volumes, variety of methodological approaches, and subfields. The following does not pretend to be a comprehensive review of either. Rather, this section discusses the major ideas within each that will enable us to appreciate fully the importance of Eisenhower's nuclear decision making during this period.

The logical starting point for any discussion of the literature on presidential power in foreign policy decision making is Richard Neustadt's *Presidential Power*. First published in 1960, this book has influenced several generations of scholars and remains the dominant work in the field.[4] Neustadt broke with the traditional approach to studying the presidency by asserting that the president's formal command power is useful only in "rare" circumstances.[5] Exercising presidential power by command is "not a method suitable for everyday employment," Neustadt wrote.[6] Rather, to exercise power successfully, a president has to rely on his persuasive abilities to convince others in the government to follow his policies.[7] The government acts, according to Neustadt, as a result of a bargaining process in which all participants enjoy some leverage because of their shared powers—a situation not unlike that existing in legislative politics.

Neustadt recognized that bargaining with presidents never takes place on a level playing field. Presidents possess the authority and status of the office, which accord them considerable advantages. Further, objects of presidential influence often know they will be dependent on the president for some favor in the future. Neustadt asserted that the president's persuasiveness is conditioned by how others in the Washington community view him. This "professional reputation" develops slowly over time through an accumulation of behavior and statements. Once established, it can be altered only marginally.[8] "Popular prestige" or public opinion is an additional variable in the Washington community's judgment of the president's professional reputation. A president should protect his popular prestige, Neustadt asserted, by seizing opportunities to "teach" the public.[9] With presidential power resting on such a tenuous foundation, Neustadt warned, the president needs to protect his personal power in his every action and decision. "He makes his personal impact by the things he says and does. Accordingly, his choices of what he should say and do, and how and when, are his means to conserve and tap the sources of his power. Alternatively, choices are the means by which he dissipates his power. The outcome, case by case, will often turn on whether he perceives his risk in power terms and takes account of what he sees before he makes his choice."[10] Given modern government's complexity, the limitations on available information, the variety of issues, and time constraints, this is a herculean task for any individual. But according to Neustadt, it can only be a solitary one.[11]

In *Presidential Power*, Neustadt described how the political environment can constrain presidents and offered prescriptions for protecting and augmenting power. Writing in the wake of strong presidents such as Franklin D. Roosevelt and Harry S. Truman and in the middle of the two Eisenhower administrations, Neustadt assumed that all presidents have activist agendas that can be achieved with the aid of others in the government. "Here is testimony that despite his 'powers' he does not *obtain results* by giving orders—or not, at any rate, merely by giving orders. He also has extraordinary status, ex officio, according to the customs of our government and politics. Here is testimony that despite his status he does not *get action* without argument. Presidential power is the power to persuade" (emphasis added).[12] For Neustadt, the business of presidents is to get action, which can be done only by persuading others to follow, not by issuing commands. But Neustadt left many questions unanswered. Will all future presidents have activist agendas? Can less activist agendas be pursued that require less bargaining or persuasion? Will presidents willingly sacrifice their power and political capital to achieve policy goals? If so, under which circumstances?

Nowhere has Neustadt's influence been greater than in Graham Allison's seminal study of foreign policy making, *Essence of Decision: Explaining the Cuban Missile Crisis*. In this book Allison developed two alternatives to the traditional "rational actor" explanation of foreign policy behavior: the organizational process model and the governmental politics model. In formulating the governmental politics model, Allison drew heavily on Neustadt[13] as well as on scholars of defense policy such as Samuel Huntington, Paul Hammond, Warner Schilling, and Roger Hilsman.[14] The governmental politics model retained the central elements of Neustadt's *Presidential Power*. This model explained foreign policy "action" as the "resultant" of a bargaining process conducted among senior policy makers. Allison's third model became best known by the phrase, "players . . . make government decisions not by a single, rational choice but by the pulling and hauling that is politics."[15] Allison had taken Neustadt's basic premise, that presidents must persuade others to do what they desire in order to be effective, and used it as the starting point for a model of decision making throughout the government bureaucracy.

Allison's elaboration of the governmental politics model included many observations that were absent from Neustadt, however. Allison emphasized the importance of organizational affiliation and position in establishing each participant's priorities, perceptions, and goals. Although a participant's perceptions of "national, domestic, and personal interests" might temper organizational influences, Allison's well-known statement "Where you stand is where you sit" indicates that, all in all, organizational factors occupied a preeminent position in determining bargaining behavior.[16] Bargaining within the executive branch over policy would proceed along issue-specific "action-channels"; the outcomes would be determined by the ability of participants to utilize their power resources. These action-channels were governed by rules that defied generalization because of their variety and uniqueness.[17] Allison also believed it inappropriate to describe policy produced by the governmental politics process as "choice." He explained that "governmental action does not presuppose government intention. The sum of behavior of representatives of a government relevant to an issue is rarely intended by any individual or group. Rather, in the typical case, separate individuals with different intentions contribute pieces to a *resultant*" (emphasis added).[18] It is equally important, as Robert Art instructs, to recognize what Allison did not take from Neustadt,

Huntington, and the others; namely, that nonbureaucratic factors such as Congress, domestic politics, and public opinion also affect outcomes.[19] Allison's third model depicted foreign policy making as the result of an executive branch bargaining process that spit out some compromise policy. Although some actors might be attuned to domestic pressures, much of the policy process remained insulated from Congress and society.

In some respects it is surprising that Allison based his governmental politics model on Neustadt's analysis. Allison made no effort to delineate the president's role and authority in either the governmental politics model or the organizational process model, even though he paid lip service to Neustadt's conception of the president as "only a 'superpower' among many lesser but considerable powers."[20] Gone was Neustadt's assertion that presidents are central in the policy process because of their special vantage point. Whereas Neustadt argued that choices have such a critical impact on success that presidents need to protect their decision authority even from their own staffs, Allison made no mention of presidential choices—and even rejected the use of "choice" to describe policy outcomes. Allison's examination of organizational constraints in both his second and third models emphasized the limits of presidential power to make and implement foreign policy.

Presidential Power and *Essence of Decision* are classics of modern political science that have dominated studies of the presidency and foreign policy for decades. Both authors were concerned with explaining how to get the machinery of government into action. Action results, the two agreed, not from commands but from persuasion and bargaining among a group of individuals sharing power within the executive branch. The two presented different perspectives on the influence of actors other than the bureaucracy, but they agreed that presidents enjoy less power to command, direct, and decide the course of government policy and activity even within the executive branch. They disagreed only on the degree of presidential constraint.

Critical examinations of Neustadt and Allison as well as subsequent scholarship in the fields of the presidency and foreign policy have raised a number of questions about this perspective. There are four issues that are crucial to the understanding of Eisenhower's role in nuclear policy making during the period of the missile gap: the assumption of an activist agenda; the balance between bargaining and command; presidential influence on the structure of policy making; and the differences arising out of the substantive nature of the policy under deliberation.

Both Neustadt and Allison assumed that the ultimate end in policy making is for the government to act. Presidents depend on others in the government to approve, support, and implement their policies. Because these other individuals have their own agendas, a president must be able to persuade these individuals to follow his lead. Consequently, bargaining arises out of the need to move the government to action. In the area of foreign policy, however, the decision-making process includes much more than just choices that prompt government action. Presidents have some, although not complete, control over their foreign policy agenda. Once an item is on the agenda, a president can decide not to act on it, either by making a formal negative decision or simply by deciding not to decide. Additionally, the absence of presidential action does not automatically empower the bureaucracy to make policy. Certain policies require presidential endorsement and involvement to be implemented or even to placed on the national agenda.

Recognition that presidents make many policy decisions that do not entail presidential action leads back to the question of the balance between bargaining and command. The politics involved in non-action decisions would seem to be closer to Neustadt's "self-executing" decisions because such decisions would require little or no presidential bargaining. Whom would a president need to persuade if he decided not to act? Inaction might not be popular with the public or Congress, but it could not be overturned easily or perhaps at all.[21] If bargaining did occur in such situations, executive officials, lobbyists, and Congress would likely beat a path to the president's door to persuade him to act. The president would enjoy virtually all of the bargaining advantages in such cases; recognizing, of course, that inaction might still bring political problems. Many of these decisions could hardly be labeled "command" decisions, since presidential commands normally connote the president's ability to use his constitutional authority to order some action without challenge. Thus, the nature of the decision can alter the bargaining dynamic in policy making in favor of the president.

Critics of Neustadt and Allison, as well as some subsequent scholars of the presidency, have argued that presidents have many opportunities to shape the structure of policy making through their control of appointments, decision-making structures, and action-channels. Robert Art, Peter Sperlich, and Stephen Krasner all assert that, through their appointment power, presidents can minimize the need to bargain.[22] Effective selection can place persons in office who identify with the president's ideology and policy goals, are beholden to him, and feel a shared responsibility for administration success and failure. Such officials would be unlikely to cling to organizational perspectives in the face of presidential pressure or to deliberately sabotage policy implementation and would feel the constitutional weight of the office. In fact, American history since the mid-1960s illustrates precisely the point made by Art, Sperlich, and Krasner: Presidents have had no trouble finding people who are willing to follow their commands blindly—even orders that violate statutes and the Constitution.

Presidential power in the policy-making process can also be accentuated through the effective and efficient organization of White House and executive branch institutions and the relations between them. This is, of course, easier said than done because each peculiar organizational arrangement has its own set of trade-offs. Presidents must find an organizational structure that strikes a balance between the competing values of maintaining presidential decision-making authority, enabling the president to gain access to important information, and allowing for the formulation of different policy alternatives, among others.

Some scholars have focused on how the president's institutional prerogatives can enhance his decision-making authority and reduce the need to bargain. Presidential choices about the organization of the Executive Office of the President (EOP) and its coordination with other bureaucratic institutions can influence the policy process in numerous ways. Institutional arrangements differ among administrations for such reasons as differences in presidential personality, administration goals, and the existing institutional structures.[23] But scholars agree that presidents usually organize their White House in one of three ways: formal, collegial, or competitive.[24] Regardless of the differences between administrations, all presidents need to organize their White House in a manner that allows them to pursue goals successfully, retains their authority to make decisions, makes the most of their individual strengths while compensating for their individual weaknesses,[25] and avoids chronic decision-making errors such as misperception and groupthink. This is, of course, no mean feat since even the most effective White House institutional arrangements can prove to be deficient under certain circumstances.[26]

Some scholars have argued that the growth of presidential institutions since the mid-1960s has transformed the presidency from a personal office into a larger, more bureaucratic organization. Terry Moe points out that presidential institution building and reorganization are powerful tools in dealing with the bureaucracy and Congress.[27] The growth of presidential institutions has resulted in greater centralization of policy decision making in the White House. But John Burke notes that this presents certain risks, such as inadequate information, presidential isolation, a more bureaucratized and less creative EOP, and increased politicization of policy.[28] Institutional growth and efficient organizational structure have provided presidents the opportunity to gain some advantage in policy making. Whether a president makes the most of it is another matter.[29]

Finally, there is a widespread recognition among scholars that presidential power in policy making varies according to the policy area. The most obvious dichotomy is that between foreign policy and domestic policy. In his famous article "The Two Presidencies," Aaron Wildavsky argued that presidents enjoy greater success in Congress in foreign policy than in domestic policy.[30] Similarly, Theodore Lowi noted differences between types of policies rather than specific issue areas.[31] Paul Light classified presidential agenda items according to the criteria of large/small and new/old.[32]

Several factors seem to be responsible for this variation in presidential power between policy areas. First, presidential power is greater in some areas, such as foreign policy, because of constitutional authority and Supreme Court decisions. Congress, the bureaucracy, and the public recognize and defer to presidential authority in such areas; effectively strengthening the president's command authority. This is not to say that a president enjoys unfettered power in any single area—that is clearly the antithesis of representative democracy—only that presidents have more latitude in some areas than in others. Second, each policy area has its own set of actors both within and outside the government who are interested in and participate in the policy process.[33] Presidential behavior with regard to policies differs, in part, according to the control these actors exercise over certain things desired by presidents, such as information, political rewards, and successful policy implementation. Third, presidential power varies across issues because of the nature of the policy. Complex issues that are beyond the expertise of the White

House or decisions that require specialized information or analysis limit presidential power because they require a dependence on others for the information that shapes policy decisions. Although policy areas can be classified in different ways, there are three that are important for this study: The constitutional power of the president in a particular policy area; the actors attentive to and active in the policy area; and the nature of the policy area.

To summarize, we noted that Neustadt and Allison argued that a president must bargain with other actors in order to produce government action in policy making. Neustadt maintained that presidential bargaining is conducted with actors throughout the Washington community who react to the president's professional reputation and public support. Allison focused on policy making within the executive branch and emphasized organizational influences. His organizational process model and his governmental politics model accorded the president no special role, authority, or influence. Criticisms of these books as well as some subsequent scholarship on the presidency have emphasized several factors that may affect presidential power in policy making in addition to those described by Neustadt and Allison. We have noted that presidents make many decisions that do not demand action or require implementation, thus reducing the need to bargain. Consequently, there may be more of a balance between bargaining and command than Neustadt or Allison admitted. The increase in EOP institutions also provides presidents with some advantages in the policy process, although utilizing them to the fullest extent requires an efficient White House organization. Finally, we recognized that presidential power varies with different issues, depending on the president's constitutional power in a policy area, the actors involved in the policy area and the nature of the policy.

We turn now to the literature on the Eisenhower presidency.[34] Scholarship on the Eisenhower administration is usually divided into two schools: the contemporary or "conventional" school and the "revisionist" school. They present starkly different views of Eisenhower, his leadership, and his policies, although some differences may be more imagined than real.

The conventional view of the Eisenhower presidency arose out of the judgments of contemporary scholars, journalists, and public officials. It depicted Eisenhower as a likeable but inactive moderate with a minimalist agenda who followed public opinion rather than leading it. Fred Greenstein, a prominent Eisenhower revisionist, described the core elements of the conventional school: "According to the conventional judgment, Eisenhower's presidency reflected the inertness and lack of skill and purpose of an aging hero who reigned more than he ruled. In this assessment, Eisenhower's professional background incapacitated him for the pull and haul of presidential leadership, because, unlike political leaders, generals can count on their orders' being obeyed."[35] With the wisdom of policy choices still debatable and access to internal administration deliberations unavailable, contemporary evaluations of Eisenhower concentrated on policy content and direction. Here they found much with which to take issue. Eisenhower's moderate leadership and unwillingness to press the nation on issues such as civil rights troubled many, particularly in the wake of the New Deal. This conventional view became more dominant in the later years of the administration as it struggled to deal with economic recession and the Soviet missile challenge. President John F. Kennedy's youthful vigor and activist agenda—as well as literature on the Kennedy administration—seemed only to reinforce the conventional view of the Eisenhower administration.

The conventional view is perhaps best exemplified by Neustadt's *Presidential Power*, in which Eisenhower is cited as an example of everything the president and the office should not be. Neustadt considered Eisenhower ill-prepared for the office, for he "had behind him the irrelevancy of an army record compiled for the most part outside Washington."[36] His experience, Neustadt noted, bred a distaste for politics and politicians. Eisenhower's inexperience in the ways of politics and his vision of what his administration should be hampered his effectiveness, according to Neustadt: "Eisenhower wanted to be President, but what he wanted from it was a far cry from what FDR had wanted. Roosevelt was a politician seeking personal power; Eisenhower was a hero seeking national unity. He came to crown a reputation, not to make one. He wanted to be arbiter, not master. His love was not for power but for duty—and for status. Naturally, the thing he did not seek he did not often find."[37] Neustadt described how Eisenhower allowed his professional reputation to decline in the year after his reelection: "He increased the insecurities attendant on supporting him and lessened the apparent risks of openly opposing him. He managed this by seeming both unsure of his objectives and unwilling to persist, for long, in any given course. His words and actions cast increasing doubt not only on his skill but on his will. Both came to be discounted in the Washington community. The impact of his influence was marked."[38] Neustadt also argued that Eisenhower structured his White House so that it was likely to fail.[39] Throughout *Presidential Power*, Neustadt compared the moderate, apolitical Eisenhower with the activist, political Roosevelt—and the former general came up short on every measure. Neustadt advocated that presidents be professional politicians with activist agendas: "The Presidency, to repeat, is not a place for amateurs. That sort of expertise can hardly be acquired without deep experience in political office. The Presidency is a place for men of politics."[40] Neustadt made clear which category applied to Eisenhower: "Desire for an amateur is not new in American politics; Wendell Willkie's instance makes that plain. Now we have had Eisenhower."[41] *Presidential Power* was as much a bible for the Eisenhower conventional wisdom as it was for activist presidents.

Numerous other political commentators in the 1950s subscribed to the "conventional" thinking about Eisenhower. One journalist observed that Eisenhower "cannot even adhere to any one position for very long because that means decision. Decision means a choice among possible alternatives and that means saying 'No' to somebody—which might offend them."[42] This moderate leadership, the journalist asserted, prevented Eisenhower from making the difficult policy decisions that were necessary. Historian Norman Graebner wrote in 1960 that Eisenhower exercised effective personal leadership which fit American "complacency" in the 1950s but gave rise to other problems. Like Neustadt, Graebner criticized Eisenhower for being apolitical and uninterested in ideas, and for allowing himself to become isolated. The result was a personally popular president who failed to lead in policy when circumstances demanded action: "The difficulty was not the President's firmness; it was the nature of his policies. It was less the decisiveness than the decisions themselves. Despite the energy behind them, Eisenhower's actions still suggested that there were no problems that good intentions would not cure."[43]

Richard H. Rovere, a political commentator for the *New Yorker* and other magazines, found similar defects in Eisenhower's presidency. He believed Eisenhower had little desire to perform the tasks of a chief executive: "Though at times he has seemed to work up a certain zest and relish for the business of being President, it is plain that most of the time the whole operational side of government has bored him."[44] Rovere concluded that Eisenhower must have found the details of governing "tedious and fatiguing" because "he at no time made much of an effort to keep up with them."[45] According to Rovere,

Eisenhower's disinterest led to the construction of a staff system that left him isolated, ignorant of details, and detached from decision making.[46] The president gave the impression of "a distressed, flustered, put-upon man" who came to the office unprepared and never grew into the job.[47]

Neustadt, Graebner, and Rovere agreed on a number of central points: the limited utility of Eisenhower's military background; his apolitical nature and lack of interest in the intricacies of governing; the formation of a White House organization that kept him uninformed and detached from decision making; his unwillingness to move beyond his moderate agenda; and the need for future presidents to be more active and assertive. The conventional view of Eisenhower dominated scholarly and popular discourse for several decades.

Eisenhower revisionists usually trace the movement away from the conventional view to works by Murray Kempton and Garry Wills, published in 1967 and 1969, respectively.[48] It would be an error to believe, as most revisionists apparently do, that the conventional view was accepted by contemporary scholars and commentators without dissent. A number of contemporary evaluations extolled the virtues rather than the vices of Eisenhower's leadership. Rovere noted that Eisenhower's foreign policy leadership was largely successful during the first term, though there were some problems.[49] The president's moderate leadership served the nation well on several occasions: "But in every crisis, he [Eisenhower] held to his own view, and it is not beyond the realm of possibility that his uncharacteristic initiative saved mankind from disaster."[50] Rovere was not entirely comfortable with this praise, however: "One hesitates to attribute political adroitness to a man who has revealed as much political ineptitude as Eisenhower, but it happens to be a fact that he has achieved, through luck or good management, a number of things that are thought to be the product of skill."[51] In a laudatory biography, Merlo Pusey argued that Eisenhower constructed an effective White House staff arrangement and achieved success in both foreign policy and economic policy during his first term.[52]

Samuel Lubell offered perhaps the most interesting contemporary analysis of the Eisenhower administration that deviated from the "conventional" wisdom. In *Revolt of the Moderates*, published in 1956, Lubell described the central "mysteries of the Eisenhower Presidency—how a general who had never run for public office before and didn't even vote until he was fifty-eight years old could be transformed, within three years, into one of the most masterful politicians in American history." Lubell recognized that Eisenhower's background and image did not conform to the traditional model for a politician. But he argued that "if . . . one defines 'politics' in its broadest sense—as the art of governing people through other people—there is little question that Eisenhower must be rated as a highly skilled professional, as compleat a political angler as ever fished the White House."

Lubell identified four elements in "the Eisenhower magic political mix" that engendered public support for the president. The first was the president's "driving determination to win" in any endeavor including politics. The second element was an idealism tempered by practicality: "This combination of an idealistic sense of duty implemented by practical means can be extraordinarily effective politically. Often, in fact, it may yield much the same end results as cynical maneuvering."[53] Eisenhower's tactics for dealing with Senator Joseph McCarthy, according to Lubell, reflected this practical idealism. It was also manifested in the president's unwillingness to make "frontal assaults upon people he disagrees with" or "rebuke" Cabinet members "openly."[54] The third element was Eisenhower's "strong sense of organization," which he applied to partisan politics. The fourth element was the president's "remarkably acute sense of public relations," which included his personal warmth and optimism. Lubell argued that Eisenhower's "political character" produced a formidable political leader: "In his own approach to politics Eisenhower has managed to combine Cromwell's zeal to win and sense of a Heaven-bound mission with a keen sensitiveness to the psychological intangibles that make for good organization and effective public inspiration." Together, these four elements constituted an effective leadership style whose essence was "the skill with which he has followed the public mood."[55] Lubell believed they explained why Eisenhower had become "Ye Compleat Political Angler."[56] But despite the contradictory interpretations by Lubell and others, the scholarly and popular view of Eisenhower coalesced around the "conventional" view and remained there throughout the 1960's.

Revisionist scholarship on the Eisenhower administration first began to appear, as already noted, in the late 1960s and early 1970s. Presidential failure in Vietnam and Watergate produced an impulse to reexamine the Eisenhower presidency that was strengthened by the opening of the administration's voluminous records at the Dwight D. Eisenhower Library in Abilene, Kansas. These archives revealed new information on virtually every aspect of Eisenhower's presidency, policies, personality, and politics. In the next two decades, scholars produced an extraordinarily large number of works, ranging from comprehensive biographies to inquiries into specific areas of foreign and domestic policy as well as countless other subjects.[57] This vast literature—lumped under the heading of "Eisenhower revisionism"—has replaced the conventional view almost completely. As Richard Immerman remarked in 1990, "Eisenhower revisionists are now a dime a dozen."[58] Any summary of the revisionist literature must be qualified or tentative because of its size and the breadth. Still, two bodies of Eisenhower scholarship stand out as being most relevant to this study: examinations of Eisenhower's leadership style and evaluations of his policies. A brief discussion of each will enable us to judge its applicability to this study.

Agreement among revisionists is greatest with regard to Eisenhower's management and leadership skills. They reject, almost unanimously, the conventional view of the apolitical Eisenhower who was uninterested and unable to govern. The best revisionist examination of Eisenhower's leadership is Fred Greenstein's *The Hidden-Hand Presidency: Eisenhower as Leader*. Greenstein argued that Eisenhower used six "strategies" that "enabled him to balance the contradictory expectations that a president be a national unifier yet nevertheless engage in the divisive exercise of political leadership."[59] Greenstein labeled the first strategy "hidden-hand leadership" in which Eisenhower used his apolitical reputation to shield his activity to achieve his goals. Second, Eisenhower relied on "the instrumental use of language" to tailor his message according to the audience and his immediate objectives. The third strategy employed by Eisenhower was an unwillingness to make his criticism personal or, in his words, to "engage personalities." Eisenhower's fourth strategy was to assess the personalities of actors prior to acting. The fifth strategy called for "the selective practice of delegation," which would allow an organization with a common objective to develop without limiting Eisenhower's decision-making authority. Greenstein argues that together, these strategies enabled Eisenhower to succeed in his sixth strategy, building and maintaining public support: "By keeping the controversial political side of the presidential role largely covert (without, however, abdicating it) and casting himself as an uncontroversial head of state, he [Eisenhower] maintained an extraordinary level of public support."

Greenstein, John Sloan, and other revisionist scholars have noted how in organizing his White House, Eisenhower established both formal and informal advisory structures and employed them in a creative and flexible way that enabled him to receive the information he needed, maintain his authority to make decisions, and cultivate cohesion within his administration.[60] This interpretation turned the "conventional" school on its head. White House organization became a means of strengthening presidential control over policy, rather than a source of presidential ineffectiveness. When revisionists such as Greenstein combined this method with Eisenhower's other management practices, they concluded that he was far more politically astute and had much more control over his administration than the conventional wisdom had allowed. Citing the new documentary evidence, these revisionists affirmed that Eisenhower was indeed "Ye Compleat Political Angler," as Lubell had argued in 1956.[61]

The lack of unanimity among Eisenhower revisionists becomes more evident when the focus moves from general assessments of White House organization and management to specific policies. We can separate revisionist studies of policy into two categories. First are the scholars whose research utilizing the newly available documents reinforces the conventional wisdom about a particular policy. These scholars can be considered "revisionists" only to the extent that they expand the empirical basis.[62] Second are the scholars whose research revises the conventional wisdom about a policy. Such policy revision can include arguments that policy choices were much wiser, that the president's objectives for the policy were different, or that some policy areas received a higher priority. These are difficult cases to make. Documents may confirm that Eisenhower was better informed or more interested in politics than had been thought, but they are less valuable for assessing the quality of policy choices. Consequently, the revisionist arguments about Eisenhower's policy choices raise some important questions. Did current perspectives influence the scholar's evaluation about Eisenhower's decisions? Does this reassessment reflect a mistaken confidence in what the new documents can reveal? Does it discount some information or perspective simply because it is old and overemphasize other data just because they are new? Does it confuse an efficient policy process with effective choices or objectives? These are important questions as we examine Eisenhower's nuclear policies during the missile gap period and place them in the context of the revisionist literature.

The literature on presidential power and the Eisenhower presidency is divided into competing schools and perspectives, but in many respects they share a common ground. The Eisenhower revisionists and the critics of Neustadt and Allison would be largely in agreement, just as those espousing the conventional view of Eisenhower would generally agree with Neustadt and Allison. The overlap between perspectives is not accidental, since the schools in agreement either developed simultaneously or provided evidence for one another. Neustadt and Allison relied in part on the theorists of the conventional school, and more recent scholars of presidential power frequently cite Eisenhower revisionists. There are even a few scholars, such as Richard Neustadt, John Burke, and Fred Greenstein, whose work is represented in both the literature on presidential power and that on the Eisenhower presidency. Nevertheless, we shall maintain the distinctions between the two bodies of literature and the various schools within each throughout this book. This will enable us to be more precise when illustrating the strengths and weaknesses of each for explaining the successes and failures of the Eisenhower administration's nuclear policies during the Missile Gap period. I do not plan to "test" each school and then to declare which schools are the "best" and which are the "worst"; I believe that each school makes an important and unique contribution to our understanding of nuclear policy during this period. It will be useful, in the remainder of this chapter, to review Eisenhower's nuclear policies before the launch of Sputnik I.

The Eisenhower Administration and Nuclear Policy, 1953–1957

When Dwight D. Eisenhower took the oath of office on January 20, 1953, he inherited from the Truman administration a set of confused policies on atomic weapons. Truman relied on U.S. nuclear superiority to support his Cold War containment policies generally and made atomic threats in both Berlin (1948) and Korea (1950 and 1951).[63] After initial post-World War II neglect, the Truman administration invested in more production facilities, which yielded a stockpile of 1,000 atomic bombs in 1953.[64] Truman and Congress established the Atomic Energy Commission (AEC) in 1946 to manage atomic weapons development and to maintain physical control over the weapons until the president approved their release to the military. These special arrangements for ensuring civilian custody of atomic weapons never translated into presidential guidance about employment strategies, however. Consequently, operational concepts for atomic weapons were left to the military, and SAC specifically.[65] But the military was uncomfortable with this situation. It feared that AEC custody, bomber aircraft range limitations and dependence on overseas bases, and a variety of other constraints would critically compromise U.S. capability to employ atomic weapons.[66]

Soviet detonation of an atomic device in 1949 and the North Korean invasion in 1950 seemed to signal that any advantage the United States gained from atomic superiority was rapidly deteriorating.[67] Some officials advocated, usually in veiled language, that the United States wage a preventive war before the Soviets could acquire a large atomic arsenal; the Truman administration responded with a buildup of conventional and atomic weapons as well as the development of a thermonuclear (hydrogen) bomb. These policies increased America's capacity to deter and defeat attacks on itself and its allies. But questions remained about the nation's long-term security in view of the Korean stalemate, the conventional force imbalance in Europe, the projected Soviet atomic arsenal, and the budgetary effects of high defense spending.

Eisenhower had played a role in the formulation and implementation of the Truman administration's security policy, having served as Army chief of staff (CSUSA), chairman of the Joint Chiefs of Staff (CJCS), and the first supreme allied commander, Europe (SACEUR). The Truman administration's domestic programs and the views of the isolationist wing of the Republican Party, led by Robert Taft, prompted Eisenhower to run for president in 1952 in order to maintain an internationalist foreign policy and a strong economy. Eisenhower attacked Truman's foreign policy throughout the campaign, promising "to go to Korea" and to replace containment with a more assertive policy of liberation or "rollback" of communism.[68] He received a mandate for these policies in the form of a landslide victory over Adlai Stevenson in the 1952 presidential election.

Eisenhower brought the lessons he had learned in the military about the importance of personnel and organizational management to his administration. His key foreign policy appointments included John Foster Dulles as secretary of state, Allen Dulles as director of central intelligence (DCI), and Charles E. Wilson as secretary of defense. Aware that these appointees might be somewhat awed by his expertise in military policy, Eisenhower made clear that he wanted them each to manage their respective bureaucracies.[69] Control over the content and direction of foreign policy would be exercised by a rejuvenated National Security Council. Whereas Truman's NSC met irregularly and sometimes without him, Eisenhower presided over weekly meetings—with a few exceptions—for the next eight years. He also increased NSC membership to include, in addition to the statutory members, the secretary of the Treasury, and the director of the Bureau of the Budget (BOB); countless others were also included for discussions on specific issues.[70] Two new NSC boards were created to help manage the increased workload. The Planning Board, consisting of the deputies of NSC members, formulated policy papers for consideration by the full NSC. The Operations Coordinating Board had responsibility for monitoring the implementation of NSC policies.[71]

To manage this expanded foreign policy-making apparatus, Eisenhower established two new White House staff offices: the Office of the Special Assistant for National Security Affairs (OSANSA) and the Office of the Staff Secretary (OSS). The special assistant for national security affairs organized the NSC agenda, chaired the meetings of the Planning Board, and managed the small NSC staff.[72] The special assistant was expected to perform these enormous tasks with minimal advocacy to ensure that the president was properly advised and to preserve his decision-making authority. The Office of the Staff Secretary kept a record of the president's meetings, both formal and informal. The work of this secretariat was important, especially after Andrew Goodpaster assumed responsibility for security issues, because it facilitated the operation of Eisenhower's informal advisory system and provided a record of it.[73] As we shall see, Eisenhower relied on informal advisory networks in making policy on very sensitive security issues and as a means of building consensus among key advisers.[74]

By the spring of 1953, the Eisenhower administration was ready to undertake a reformulation of national security policy. Campaign rhetoric and a decision to cut Truman's defense budget appropriations for fiscal year 1954 from $41 billion to $36 billion effectively precluded a strategy that continued this level of defense spending, much less one that increased it.[75] In early May 1953, Eisenhower ordered the formation of Project Solarium in which separate study groups would examine three alternative courses for national security policy: continuing the Truman administration's containment policy; delineating American national interests and deterring Soviet aggression in those areas with the threat of force; and policy of military, economic, and covert force designed to reduce and eventually eliminate Soviet control.[76] The three Project Solarium studies were presented to Eisenhower and the NSC on July 16, 1953, and became the basis for the development of a new strategy despite a number of irreconcilable differences among the three alternatives.[77] Over the next few months, the administration debated the contours of a new policy that would balance military requirements with economic costs. In October 1953, Eisenhower approved NSC 162/2, a top secret policy paper on "Basic National Security Policy" (BNSP). Eventually dubbed "the New Look," the paper called for a reduction in expenditures for security programs, to be achieved by emphasizing alliances, covert operations, propaganda, and, most importantly, nuclear weapons.[78] Nuclear weapons were "indispensable for U.S. security" and would be a deterrent against Soviet aggression if a "massive atomic capability" was maintained. Paragraph 39-b stated: "In the event of hostilities, the United States will consider nuclear weapons to be as available for use as other munitions."[79] The New Look's emphasis on nuclear weapons countered the budgetary, planning, and custody constraints imposed by the Truman administration.[80]

The first full public enunciation of the New Look came on January 12, 1954, when John Foster Dulles addressed the Council on Foreign Relations in New York City.[81] "The way to deter aggression," he told the audience, "is for the free community to be willing and able to respond vigorously at places and with means of its own choosing."[82] The bellicose speech shocked many, a reaction that led Eisenhower and Dulles to clarify several times that atomic weapons would not be employed indiscriminately in any conflict with the Communists. The New Look, especially the "massive retaliation" aspect, was attacked on a number of grounds, including its credibility as a deterrent, its applicability to "peripheral" areas such as Korea and Vietnam, and the dangers of escalation to a general nuclear war.[83] But the Eisenhower administration's national security strategy continued to be based on massive retaliation despite this criticism. Subsequent NSC papers on BNSP (NSCs 5440, 5501, 5602/1, and 5707/8) stressed the maintenance of U.S. deterrent capabilities and the continuing integration of nuclear and conventional weapons, and stated that nuclear weapons—as well as chemical and bacteriological weapons—would be considered for use in any type of conflict.[84]

In the four years between the issuance of NSC 162/2 and the launch of Sputnik I, the Eisenhower administration expanded the nation's nuclear stockpile and delivery capabilities in an effort to maintain U.S. superiority over the Soviet Union. Successful development of the hydrogen bomb made nuclear weapons far more plentiful and enabled the United States to design nuclear weapons for a range of military missions in addition to large-scale attacks on targets within the Soviet Union. The nuclear stockpile grew from 1,000 in 1953 to 2,110 in 1955 and 5,420 in 1957. Total megatonnage increased from 154 MT in 1955 to 16,300 MT in 1957.[85] The Eisenhower administration increased delivery capabilities by purchasing 956 B-47 medium-range jet bombers and 243 B-52 bombers; both types of aircraft represented significant qualitative advances. Overall, SAC bomber forces increased from 762 aircraft in 1953 to 1,655 aircraft in 1957.[86] The Eisenhower administration also approved research and development programs for many new delivery systems, including the B-58 medium-range bomber, the B-70 strategic bomber, a nuclear-powered aircraft, two intermediate-range ballistic missiles (IRBMs), the Polaris submarine–launched ballistic missile (SLBM) system, and an intercontinental ballistic missile.[87] But these programs encountered numerous problems that delayed their progress, such as scientific and technical obstacles, rivalry among the branches of the armed services, and competition for research and development funds that were very limited because of budget ceilings.

The Eisenhower administration's implementation of the New Look policy and the estimated increase in Soviet nuclear capabilities precipitated other changes in U.S. nuclear policy as well. Truman's policy of ensuring civilian custody of nuclear weapons gradually eroded as the Eisenhower administration transferred larger numbers of nuclear weapons to the military.[88] This action was dictated in part by the development of tactical nuclear weapons for such missions as air defense and European battlefield employment, which required immediate employment of nuclear weapons if they were to be decisive. But AEC custody might prevent their effective use in such circumstances. Therefore, in 1956, President Eisenhower further eased the restrictions on nuclear weapons; he approved procedures whereby the seven U.S. commanders in chief could, by declaring a "Defense Emergency," gain custody of nuclear weapons without having to obtain the president's authorization.[89]

While the Eisenhower administration undertook these measures, intelligence estimates indicated growth in Soviet nuclear capabilities as well. In a report issued in May 1953, the NSC estimated that the Soviet atomic stockpile consisted of 120 bombs and would likely increase to 300 by 1955.[90] The successful Soviet test of a hydrogen bomb in August 1953 resulted in an upward revision of intelligence estimates on the grounds that the new weapon could produce a large yield with less nuclear material. In 1954, American intelligence estimated the total yield of the Soviet nuclear stockpile to be 25 MT; yet only a year before, it predicted the Soviets would not reach that level until 1957—with an increase to 172 MT in five years.[91] According to a 1955 a National Intelligence Estimate (NIE), the number of Soviet nuclear bombs would probably increase from the then current level of 490 to approximately 1,250 by 1958.[92] Expectations that this trend would continue were confirmed by a 1956 CIA estimate of an increase in Soviet plutonium production.[93] This projection gave rise to similar estimates, such as a 1957 prediction that the Soviet atomic stockpile would increase from 850 bombs in 1957 to 3,000–4,000 bombs in 1961.[94]

Intelligence officials were also becoming increasingly concerned about the Soviet capability to deliver nuclear weapons. In 1953, aircraft presented the only means (with the exception of some unconventional operation) by which the Soviet Union could launch a nuclear strike against the United States. At that time, U.S. intelligence estimated that the Soviet bomber force consisted of 1,000 TU-4 medium-range propeller aircraft, comparable to the American B-29. This force could threaten the United States with "one-way" strikes, but operational limitations made it an unformidable first-strike weapon.[95] Still, U.S. intelligence officials believed the Soviets would introduce and deploy 180 heavy bombers with a much longer range and 120 medium-range jet bombers within the next two years.[96] Intelligence estimates of Soviet delivery capabilities increased significantly following the successful Soviet hydrogen bomb test in 1953. Estimates in 1954 of the deployment of TU-4 aircraft in that year were larger than the 1953 NIE projection of their deployment in 1957. These reports also predicted that the Soviets would deploy new bomber aircraft more rapidly than previously estimated. On June 7, 1954, NIE 11-5-54 projected the Soviet medium-range jet bomber force (the TU-39 "Badger") would number approximately 120 in 1955 (not 50 as had been estimated a year earlier) and 600 in 1959. Two subsequent revisions raised the estimate for 1955 to 200 and the number for 1959 increased first to 600–900 and then to 1,050. The NIE predicted that the Soviets would shift to heavy jet bombers and would deploy 100 of them in 1959, although this estimate too was later increased to 250.[97] Intelligence reports speculated that the Soviets might deploy a medium range missile by 1956 and an ICBM by 1959 or 1960. In less than two years, U.S. intelligence had elevated the estimates of the Soviet bomber threat just as it had those of Soviet nuclear weapons development.

The pressure to produce higher Soviet bomber estimates continued in spring 1955, when U.S. officials counted 11 heavy jet bombers ("Bison") at rehearsals for Moscow's May Day parade. Washington thought this indicated the Soviets had been more successful in developing heavy jet bombers and could produce them more rapidly than estimates had indicated.[98] Less than two weeks after the parade, NIE 11-3-55 reported that, although the Soviet bomber force would remain at approximately 1,400 aircraft, over the next five years the Soviets would retire the 1,160 antiquated TU-4s and replace them with new, mostly jet, medium-range and heavy bombers.[99] The Eisenhower administration suddenly realized this might mean the emergence of a Soviet numerical advantage in advanced bomber aircraft. This so-called bomber gap prompted the administration to call for accelerated production of the B-52. It also precipitated a major public debate between members of the administration and some congressional Democrats over whether the New Look's fiscal controls had given the Soviets an opportunity to gain ground and perhaps even to move ahead of the United States.[100]

As the administration battled the political fallout from the bomber gap, concern rose over Soviet ballistic missile research and development. Intelligence estimates indicated that the Soviets would not be able to deploy "militarily significant quantities" of ICBMs until 1960–1965, but Pentagon officials worried that the Soviets already had a two-year lead over the United States in missile development.[101] The administration knew there would be serious political and military ramifications if the Soviets were the first to develop an ICBM.[102]

Fears about the bomber gap diminished in 1956 and 1957 when intelligence officials failed to discover new bomber deployments of the magnitude predicted. Information gathered from secret U-2 spy planes that began periodic overflights of the Soviet Union on July 4, 1956, was critical in this regard.[103] Eventually, officials estimated Soviet heavy bomber deployments at 125 aircraft, not 700 as had been estimated in earlier NIEs.[104] Intelligence officials also followed Soviet missile progress by monitoring missile tests. In 1956, an NIE predicted that within three years, the Soviets would have operational medium-range missiles armed with nuclear warheads and capable of striking most U.S. overseas bases. Furthermore, the Soviets would have an ICBM capability in 1960–1961 that would "create a direct and dangerous threat" to the United States.[105]

As a result of the inflated intelligence estimates of the magnitude of the Soviet nuclear stockpile and delivery capabilities in the pre-Sputnik period, administration officials became increasingly aware of the vulnerability of the population and the military to a nuclear attack, despite U.S. nuclear superiority. A 1953 NSC report estimated the U.S. casualties resulting from a Soviet nuclear attack to be 9 million and predicted that the number would increase to 12.5 million casualties in 1955.[106] Three reports made to President Eisenhower in 1956 showed how the estimates of the casualties resulting from a Soviet nuclear attack increased to mind-numbing levels: 65 percent of the U.S. population killed or injured; 50 million killed immediately; and 40 percent killed.[107] Officials of both the Truman and Eisenhower administrations had been frustrated by the complexity, enormous costs, and ineffectiveness of programs for defending the continental United States against a Soviet bomber attack. Because of their generally poor record, the Eisenhower administration had subordinated defense and warning programs to offensive programs when determining budget priorities.[108] This decision had the unfortunate effect of denying missile warning programs adequate funds for research and development in the pre-Sputnik period.

The Eisenhower administration worried that U.S. strategic nuclear forces might become so vulnerable to a Soviet nuclear strike that they would be unable to retaliate. A reduction in the vulnerability of U.S. strategic nuclear forces required adequate intelligence, warning, and protection, both active and passive. Persuasive arguments were made that America's numerical advantage of these forces would not deter the Soviets if vulnerability prevented such forces from retaliating. In 1955, the Technical Capabilities Panel (the Killian panel) formulated a comprehensive program for protecting U.S. strategic forces from attack, improving intelligence, and ensuring that there would be adequate warning of a Soviet attack. In a 1956 report, the NSC Net Evaluation Subcommittee (NESC) emphasized that "if the United States should fail to maintain adequate alert nuclear forces that cannot be destroyed by surprise attack, the USSR by a nuclear attack on the continental United States will emerge as the dominant world power in 24 hours."[109] The Eisenhower administration approved a variety of measures designed to ensure that SAC could retaliate after a bomber attack. They included expanding SAC forces and radar warning networks, decreasing SAC reaction time through alert procedures, and dispersing SAC bombers to more airfields.[110] But these efforts would not necessarily protect U.S. forces if a Soviet ballistic missile threat materialized.[111]

From 1953 to 1957, the Eisenhower administration relied on nuclear weapons to meet its security commitments and maintained stable defense expenditures. Nuclear stockpiles and delivery capabilities increased while policy changes endorsed the employment of nuclear weapons in any type of conflict. But NIE projections of large Soviet nuclear forces led some officials to question the credibility of massive retaliation, especially for extended deterrence. The willingness of the Eisenhower administration to consider using nuclear weapons in Korea, Vietnam, and the Taiwan Straits increasingly became a cause for alarm. These issues came to a peak in the Eisenhower administration's last three years after the Soviets launched Sputnik I in October 1957.

Soviet development of strategic missiles posed three potential problems for the Eisenhower administration. First, such missiles would make U.S. strategic nuclear forces vulnerable to a surprise attack—that is, weaken U.S. nuclear deterrence. Second, the existence of a Soviet strategic missile force would raise doubts about the utility of a policy of extended deterrence for the defense of Western Europe. Third, the threat of Soviet missiles would necessitate changes in American behavior in the event of a nuclear crisis, since the United States could no longer be certain that it alone would determine when, where, and how nuclear escalation would occur. This book examines how the Eisenhower administration addressed the first of these three problems. Minimizing the vulnerability of U.S. strategic forces to Soviet strategic missiles became the central security concern of Eisenhower and his aides, overshadowing all other security issues. Strategic deterrence was the bedrock of the administration's national security strategy. If it was undermined, other security policies, including extended deterrence, and perhaps Americans' resolve in a crisis would also be.

The next four chapters focus on the policies developed by the Eisenhower administration in response to the perceived vulnerability that would be created by a Soviet missile force. Chapter 2 reviews U.S. intelligence assessments of Soviet missile programs; combined with an inadequate warning system and SAC bomber deployments, they led to worst-case scenarios of vulnerability. Chapter 3 examines the formulation of national strategy and nuclear strategy. Eisenhower remained unwilling to abandon a policy of massive retaliation but became frustrated by the way it was manifested in nuclear strategy. Chapter 4 evaluates Eisenhower's ability to control defense spending in the face of mounting pressure for increases by the armed services, Congress, the press, and academic strategists. Chapter 5 explores how the Eisenhower administration's decisions about which weapons to develop and the size and manner of deployments determined the U.S. strategic nuclear force posture. These four areas constitute the essence of the Eisenhower administration's response to the Soviet missile threat. Organizing the book in this way may seem somewhat artificial in that it de-emphasizes the overlap between the different areas of strategic nuclear policy. I employ these distinctions because doing so allows a clearer discussion of each policy area and enables the distinctive aspects of each to be highlighted.

The other two problems created by the threat of Soviet missiles—extended deterrence and nuclear crisis management—are addressed periodically in the next four chapters. The declassified record of the missile gap period allows for some general observations about these two problems.

Soviet development of medium and long range missiles created problems for extended deterrence in Europe which Eisenhower struggled with, as did his successors until the denuclearization of Europe began in 1987, following the signing of the Intermediate-Range Nuclear Forces (INF) Treaty. Eisenhower and subsequent presidents feared that the development of Soviet nuclear weapons would undermine the cohesion of the Atlantic Alliance and weaken U.S. political leadership in Europe. As early as 1955, an NIE warned about problems of allied support in the event of a nuclear crisis:

> If a general war appeared imminent or actually occurred, [allied] policies would depend in large measure on the course of events. . . . Some governments might estimate that full-scale nuclear war between the US and the USSR would end with complete or near complete destruction of the war-making potential of both powers, and therefore that neutrality might be a safe and profitable position. If events developed in such a way as to confront governments with a clear and immediate choice between nuclear devastation and neutrality, we believe that practically all would choose neutrality.[112]

In early 1957, a NSC Planning Board report asserted that increased superpower nuclear capabilities and the risk of nuclear escalation presented U.S. allies with three alternatives: "Our allies will increasingly (1) weigh the added security of U.S. alliances against what they may regard as the increased risk of associating with the U.S., (2) be susceptible to Soviet nuclear threats, (3) seek nuclear weapons in order to pursue their own interests."[113]

The Eisenhower administration's efforts to strengthen extended deterrence in Europe during the missile gap period concentrated on continuing the "nuclearization" of NATO, which had begun in 1950–1951.[114] It negotiated with Great Britain, Turkey, and Italy for the deployment of 105 IRBMs. This stopgap measure suffered from a variety of problems including poor coordination between the United States and its allies and the technological weaknesses of the weapons.[115] Perhaps most important, the restrictions imposed by the Atomic Energy Act on the transfer of nuclear weapons and information to other countries created an awkward situation: the host countries owned the missiles but the United States owned the actual nuclear warhead.[116] Long an opponent of any restrictions on U.S. nuclear weapons cooperation with its allies, Eisenhower contemplated ways to improve "nuclear sharing" among NATO member nations but without result. Although he accepted additional U.S. nuclear solutions for European defense, Eisenhower rejected any changes in national strategy or budgeting that might increase the U.S. conventional commitment to Europe (see Chapters 3 and 4).[117] The Eisenhower administration's policies with regard to extended deterrence set the stage for the Kennedy administration's flexible response strategy and the debates with NATO countries over independent nuclear deterrents. Eisenhower was the first president to struggle with the question of defending Western Europe when both it and the United States were vulnerable to a Soviet nuclear attack.

The Eisenhower administration faced two nuclear crises during the missile gap period: The second Taiwan Straits crisis (1958) and the Berlin Deadline (1958–1959). In each case, a communist nation challenged U.S. and Allied authority in a geographic area that had been the site of a previous nuclear crisis. Several aspects of these two nuclear crises are of interest for the purposes of this study. Eisenhower's style of managing nuclear crises changed little after Sputnik; it was remarkably consistent with his behavior before Sputnik. Eisenhower's management style in the nuclear crises included keeping the military's options open for the use of nuclear weapons, delegating little authority to military commanders in the field, and searching for solutions short of armed conflict. This style assumed the continuation of U.S. nuclear superiority. Eisenhower remained confident about his ability to manage nuclear crises, but his style frustrated the military, troubled journalists and the public, and has confounded scholars.[118] Most important for this study, the consistency in Eisenhower's style indicates that concerns about growing U.S. vulnerability to a Soviet first strike did not affect his nuclear crisis behavior. This is not as surprising as it might seem, since U.S. intelligence usually estimated the danger period to be more than eighteen months into the future. Thus, Eisenhower could be confident about the nation's deterrent capabilities during a crisis while worrying about them over the long term.

[2]

Soviet Missiles, Warning, and Perceptions of Vulnerability

This chapter examines American perceptions of U.S. vulnerability during the missile gap period. It focuses on intelligence estimates of Soviet missile strength, systems for warning of a Soviet attack, and projections of SAC vulnerability. The first section concerns intelligence estimates of Soviet production of long-range missiles—both ICBMs and SLBMs—during this period. From 1957 to 1960, intelligence officials consistently placed the emergence of a large Soviet ICBM force farther in the future after the initial post-Sputnik alarms. Intelligence estimates of future Soviet ICBM quality and quantity progressively increased: from a large number (±500) of crude missiles to a medium-sized force (±300 to 400) of powerful, semiaccurate, and reliable missiles. Intelligence projections consistently underestimated Soviet SLBM quality and force size. Concern about the Soviet SLBM force rose just as the Soviet ICBM threat began to decline. Inaccuracies in intelligence estimates can be attributed to a weak data base, erroneous assumptions in analysis, and poor coordination within the intelligence community. Charges of systematic bias made by some scholars are not supported because of these different inaccuracies in ICBM and SLBM estimates.

Warning systems for a Soviet attack against the United States are the subject of the second section. By the time the Soviets had launched Sputnik I, several systems had been constructed to warn of a bomber attack, but they had numerous problems that undermined their effectiveness. The United States did not begin to construct an ICBM warning system until after Sputnik I and did not begin operating it until 1961. The United States did not even bother to build an SLBM warning system, relying instead on antisubmarine warfare (ASW) and surveillance of Soviet submarines, although neither method was foolproof.

The third section illustrates how inflated intelligence estimates of Soviet missile programs and the problems of warning systems translated into worst-case attack scenarios. This perceived vulnerability persisted even after estimates had advanced Soviet ICBM deployments into the future because of the advent of Soviet SLBMs, qualitative estimates of Soviet ICBMs, and the length of time required to diversify nuclear forces and build warning systems. Eisenhower administration officials knew this information, made policy in response to it, and passed it on to the Kennedy administration.

The conventional wisdom concerning intelligence and perceived vulnerability during the missile gap period usually blames bureaucratic interests in the Air Force and the CIA for overestimating the Soviet missile program. This perspective rejects fears of strategic vulnerability, usually citing intelligence reports supplied by U-2 aircraft. But this chapter shows that the situation was actually much more complex and hence more problematic for civilian and military leaders. Intelligence information available in the pre-reconnaissance satellite period proved to be insufficient quality or quantity to satisfy decision makers. Worst-case scenarios concerning the nation's vulnerability were generated when the missile estimates were combined with evaluations of U.S. warning systems. Recently available declassified documents have yielded new insights into the complex dynamic of this perceived vulnerability.

Intelligence Estimates of Soviet Missile Capabilities

The Eisenhower administration's efforts to estimate Soviet missile progress throughout the missile gap period were made difficult by the nature of intelligence data on Soviet missiles. Most intelligence concerning these missiles came either from electronic listening posts on the periphery of the Soviet Union or from U-2 reconnaissance flights. These sources enabled the United States to monitor some aspects of the Soviet missile program more closely than others. Intelligence officials knew the Soviet test flight program best; they had analyzed both individual tests and the program's overall efficiency. A close reading of the documentary evidence indicates that the United States possessed less information on the quality and reliability of Soviet missile designs and components. Even more important, the United States needed to determine Soviet intentions with regard to the scope and objectives of Soviet missile programs, especially at this early stage. The first indication of Soviet ICBM deployments would be the construction of missile bases (which preceded deployments by approximately eighteen months), unless the CIA could gain access to information about Khrushchev's private deliberations. *No* intelligence existed that could confirm or disconfirm Soviet deployments two or more years in the future unless the projection was so large that such a deployment would be physically impossible. Lacking more complete data, intelligence officials overrelied on the testing data and applied it to reach conclusions about Soviet missile quality and deployments.

In the wake of Sputnik I, Eisenhower administration officials reassured the public about national security even as they tried to determine Sputnik's military significance. Eisenhower held numerous meetings on the Soviet missile programs with Defense Department officials and members of the NSC. The meetings confirmed that Soviet programs surpassed those of the United States in a number of important respects.[1] In 1952, the Soviets had built a rocket motor capable of approximately 200,000 pounds of thrust—twice as powerful as any U.S. rocket available in 1957. (The thrust of U.S. rocket motors ranged from 27,000 pounds to 150,000 pounds.[2]) Tests of this prototype had provided the scientific information necessary for the Sputnik launches.[3] The CIA intelligence indicated that more than 300 flight tests of missiles with ranges of from 75 NM to 950 NM were held at the Kapustin Yar complex. In the summer of 1957, the Soviets held seven flight tests of their 950 NM missile as well as initial ICBM tests. Their skill and organization enabled them to conduct twenty-two missile flight tests in one month and four in one day. Director of Central Intelligence (DCI) Allen Dulles told the NSC that the Soviets' test of a high-yield thermonuclear weapon, their claim of having held ICBM tests, and the launch of Sputnik I formed a propaganda "trilogy." Intelligence reports revealed that fewer heavy bombers had been deployed than projected, Dulles stated, perhaps indicating that "the Soviets are in the process of de-emphasizing the role of heavy bombers," as Khrushchev claimed.[4]

By late October, intelligence reports warning of a national crisis began circulating within the Eisenhower administration. On October 23, 1957, a board of CIA civilian consultants informed DCI Allen Dulles that the United States lagged behind the U.S.S.R. by "two to three years." The consultants, who had been following the progress of Soviet guided missiles for Dulles and the CIA since 1954, concluded that "the data proves beyond question that the Russians have an orderly and progressive program which is being prosecuted in an aggressive and intelligent manner. The program does not appear to us to be of a 'crash' nature but rather one that has been thoroughly thought out and followed for years. One of the most disturbing features revealed is the high level of Soviet competence in achieving their planned goals." The consultants cited Soviet missile reliability, accuracy at ranges up to 650 NM, and an engine thrust sufficient to support their conclusion. With production of a Soviet ICBM "nearly" a reality, they projected deployment of a dozen Soviet ICBMs by the end of 1958. Based on this evidence, the consultants warned Dulles that the United States "is in a period of grave national emergency."[5]

In early November, the Gaither panel endorsed the "national emergency" hypothesis when it presented its report to President Eisenhower. The report noted that the Soviet lead in rocketry would provide the U.S.S.R. with an ICBM advantage in 1959 or 1960 that would continue for about two years. The Gaither Panel did not include numerical estimates of Soviet ICBM programs but simply stated that "The U.S.S.R. will probably achieve a significant ICBM delivery capability with megaton warheads by 1959." Although the panel never defined "significant," it concluded that "this appears to be a very critical period for the U.S."[6] In a private meeting with President Eisenhower, H. Rowan Gaither presented an even darker assessment: "The evidence is overwhelming that the U.S.S.R.'s intentions are expansionist—in the real global sense—and that her great efforts to build a military power go beyond any defensive concepts." He emphasized the rapid expansion of the Soviet military power, particularly the nuclear stockpile, thought to number 1,500 weapons. Gaither warned that if U.S. defense policy remained unchanged, there might be a nuclear war between the superpowers that could result in casualties of 50 percent or more.[7] Thus, the ambiguous assessments and overarching conclusions of the Gaither panel fueled fears concerning the existence of a missile gap. The panel did not create the missile gap or "national emergency" concepts, but its report was essential to their circulation beyond the intelligence community and, ultimately, outside the administration.

While the CIA civilian consultants and the members of the Gaither panel delivered their analyses, the Office of National Estimates (ONE) of the CIA finished its annual NIE on Soviet military programs. NIE 11-4-57, issued on November 12, 1957, reflected information received on the successful Soviet ICBM test launches in the previous summer and the launch of Sputnik I.

> In light of this and other new evidence, we have re-examined our previous estimate of Soviet ICBM development, and have tentatively advanced from 1960–61 to 1959 the probable date when a few (say, ten) prototype missiles of 5,500 nautical miles (. . .) range could first be available for operational use. . . . Early success of any phase of the test program, or relaxed accuracy and reliability requirements, could advance the date of availability.

Soviet achievement of this deployment date required an aggressive flight-testing program, a circular error probable (CEP) of only 5 NM, and initial deployments of prototype rather than "series-produced" ICBMs. According to the NIE, Soviet bombers numbered approximately 1,500, although the replacement of old aircraft with long-range jets had been slowed. The NIE anticipated continued Soviet development of an IRBM with a range of 1,000 NM with possible deployments in 1958. NIE 11-4-57 included an evaluation of Soviet SLBM programs, but this section seems to have received very little attention at the time. The report stated: "we believe the Soviet surface-to-surface program also includes submarine-launched missiles" and that the Soviet Union might even possess an operational supersonic cruise missile for submarine launch with a range of 500 NM.[8]

National Intelligence Estimates generally carried great authority during the Eisenhower years due to their wide distribution throughout the national security bureaucracy and because administration policies and public statements were frequently based on them.[9] This was not true of NIE 11-4-57, however. Intense scrutiny of Sputnik data and other information continued throughout the drafting of NIE 11-4-57 and its dissemination. Not surprisingly, new information and analyses soon superseded the conclusions contained in this NIE and resulted in more somber estimates.

Two weeks after the issuance of NIE 11-4-57, DCI Allen Dulles and Assistant DCI for Scientific Intelligence Herbert ("Pete") Scoville presented a much different assessment in top secret testimony before the Preparedness Investigating Subcommittee of the Senate Armed

Services Committee. The launch of Sputnik I and II, they told the senators, put ICBM requirements for range (5,000–6,000 NM) and throw weight (2,000 pounds) within the Soviets' reach. Even though the Soviets still faced guidance and reentry difficulties, Dulles and Scoville predicted that the first Soviet deployment of ten prototype ICBMs would take place between 1958 and 1960. This estimate yielded additional projections: "One hundred ICBMs could be available [to the Soviet Union] for operational use between the middle of 1959 and late 1960 and five hundred ICBMs could be available for operational use between the middle of 1960 and 1962. [The] CIA at the present time believes that the earlier dates are the more likely but the whole problem is under urgent review at this moment." Dulles and Scoville also noted that other Soviet missiles might threaten the United States in the future. For example, each of the 250 long-range Soviet submarines could carry two cruise missiles. But Dulles admitted that "we have not yet confirmed the existence of such Soviet submarines, but we have received reports from widely separated areas describing subs with suspicious looking topside installations." The cruise missiles would not be replaced by ballistic missiles, Scoville reported, until the mid-1960s.[10]

On December 17, 1957, the ONE issued a special NIE (SNIE) devoted exclusively to Soviet ICBMs. SNIE 11-10-57 affirmed the higher estimates stated by Dulles and Scoville in their testimony as the official intelligence estimate of Soviet missile deployment, replacing the relevant sections of NIE 11-4-57. According to the new estimate, initial ICBM reliability would be 50 percent, but the figure was expected to increase to 70 percent after the Soviets had deployed several hundred ICBMs.[11]

In the two months after the launch of Sputnik I, intelligence assessments became increasingly pessimistic. The testimony of Dulles and Scoville was the first indication that the administration acknowledged the possibility of a missile gap between the superpowers in the early 1960s, although the testimony did not create a stir at the time. From this point on, intelligence estimates consistently downgraded the Soviet ICBM threat, placing it farther into the future. But the damage had already been done in the first two months after Sputnik. President Eisenhower and the CIA spent the next three years trying to repair it.

Soviet ICBM testing followed an erratic course in 1958, which was reflected in the NIEs. Intelligence analysts monitored Soviet ICBM progress by studying the amount of flight-testing observed and then extrapolating the data to determine an initial deployment date as well as a probable date for the deployment of a large number (perhaps 100) of operational ICBMs. By mid-1958, the CIA counted tests of almost 300 medium-range ballistic missiles (with ranges of up to 1,000 NM), 12 intermediate-range ballistic missiles (1,100 to 1,500 NM), and 6 long-range ballistic missiles (3,500 NM). However, radar facilities in Turkey and the small number of U-2 flights failed to detect any Soviet test flights of ICBMs (with ranges of more than 3,500 NM) after late May.[12] Air Force Intelligence and the CIA Office of Scientific Intelligence (CIA-OSI) disagreed over the meaning of the absence of tests. The Air Force claimed the Soviets had completed testing and had already entered the deployment phase; CIA-OSI countered that the testing moratorium resulted from unresolved technical problems.[13]

This bureaucratic dispute came at an inopportune time for Allen Dulles. A revised NIE, delivered on May 20, 1958, endorsed the CIA-OSI position and projected Soviet achievement of a ten-missile initial operational capability (IOC) in 1959 rather than in mid-1958 as had been predicted in the previous NIEs.[14] Dulles expected that the Soviets could have a force of 500 operational ICBMs two years after these initial deployments (mid-1961).[15] When he conveyed this revised estimate to a Senate committee in the summer of 1958, he inadvertently fueled a political controversy over Soviet missile development.

Senator Stuart Symington (D-Mo.), the first secretary of the Air Force and former chairman of the National Security Resources Board, challenged the revised estimate. Throughout the 1950s, Symington's commitment to a strong Air Force had clashed with the Eisenhower administration's measured approach to missile development and defense programs, especially during the bomber gap period. Symington became suspicious when he heard the new intelligence estimates. Through a former assistant, Thomas Lanphier (who at the time was an assistant to the president of Convair—manufacturer of the Atlas ICBM), he received reports of larger numbers of Soviet missile tests that contradicted the official position of the CIA.[16] If correct, this alternative intelligence on Soviet tests meant that they already had enough reliable data to begin ICBM production and deployment.

Twice that summer, Symington and Lanphier confronted DCI Dulles with their alternative intelligence. Dissatisfied with Dulles's doubts that "Lanphier's figures could be backed up in any intelligence component in the Government,"[17] Symington appealed directly to Eisenhower. In a private meeting with the president, he detailed the implications of the disagreement between the CIA and his intelligence information.

> The United States plans to have 24 operational ICBM's by 1960; and 120–130 by 1962. In the same time period the CIA estimates the Soviets will have 500 ICBM's by 1960 or 1961. Based on these accepted figures alone, we believe our currently planned defense programs are insufficient to meet the threat which the CIA estimates the Soviets will pose by 1961. . . . [If our contention about Soviet missile testing is true] it is clear that our planned defense programs are even more insufficient. May we respectfully present the fact that you have said many times we (*sic*) should not underestimate a possible enemy. Based on the information outlined, however, we believe our national intelligence system is underestimating the enemy's current and future ballistic missile capability.[18]

Unpersuaded, the president reminded the senator of the problems that had been caused by inaccurate intelligence during the bomber gap period—charges Symington had made several years earlier—as well as the danger of premature weapons development. Eisenhower downplayed the reports concerning Soviet missile test flights and expressed his continued confidence in the nation's strategic programs.[19]

Eisenhower reassured Symington and encouraged him to discuss the issue in greater depth with Defense Secretary McElroy, but the president was sufficiently concerned about the accusations to forward the senator's letter to Allen Dulles for explanation. Dulles replied to Eisenhower that the CIA could find no evidence of Soviet construction of new launch sites nor could he uncover "any effort within the Intelligence Community to suppress evidence or to prevent the fullest analysis of the views of any competent person with information to contribute on this vital subject." He also reminded the president of the uncertainty involved in CIA projections concerning Soviet long-range missiles: "There is no direct evidence on Soviet plans for quantity production of ICBMs and we recognize that reasonable men might differ as to these figures; in fact, it is possible that the Soviet [*sic*] themselves do not have as yet fixed plans in this regard."[20]

With the release of the annual National Intelligence Estimate on the subject (NIE 11-4-58) fast approaching, Allen Dulles hoped that a review by the Guided Missile Intelligence Committee of the United States

Intelligence Board could placate Symington and convince him as well as Air Force officials to support the downward estimates. He asked the committee to "re-examine the Soviet ICBM development program and to reaffirm or recommend modification of our existing estimate." He also asked the committee to answer a wide range of very specific questions and report back to him and the Board of National Estimates by October 31, 1958.[21]

With the committee's report and the new NIE in hand, Dulles briefed Symington and Lanphier in mid-December.[22] Dulles reviewed Symington's charges concerning inaccurate intelligence and presented the gist of the new NIE. In the area of missile test firings, the committee had found little evidence to substantiate Symington's claims. Soviet patterns of testing missiles with ranges greater than 700 NM displayed an uncharacteristic irregularity. The number of test firings of long-range missiles (ranges exceeding 3,500 NM) had not increased since May, when the figure was 6—although the Soviets suffered two test failures in July. Medium- and intermediate-range missiles, Dulles reported, remained the areas of greatest Soviet activity. In one three-month period in 1957, the Soviet Union conducted its first 8 flight tests of missiles with a range of 1,000 NM. Although the Soviets held only 3 such tests in the next twelve months, 6 tests followed in the last four months of 1958, bringing the total number of tests to 17. Since August 1, he informed Symington, the Soviet Union had flight-tested 40 missiles with ranges of approximately 700 NM; it had conducted almost 400 flight tests of missiles in this range since 1953.[23] In the year since the launch of Sputnik, I, the Soviet Union had conducted approximately 125 tests of guided missiles.

According to Dulles, a surprised Symington responded to the CIA intelligence by commenting that "on the basis of these figures it looked as if the Soviets were slackening up in their ICBM development." When Scoville posited that perhaps the Soviets had gathered sufficient information to begin deployment, Symington replied that "they hadn't fired enough to get the data . . . and . . . if our evidence was right the Soviets had been way ahead of us but were slackening up." "Maybe they had gone back to the drawing board to fix something," added Air Force Chief of Staff for Intelligence Major General James Walsh. The Guided Missile Intelligence Committee, Dulles said, had been unable to find evidence substantiating his conclusion that the Soviets had held flight tests of ICBMs armed with nuclear warheads or had begun new ICBM launch site construction. He reiterated the change in the anticipated deployment dates that was to be included in NIE 11-4-58. When Symington questioned this change, Howard Stoertz of the CIA replied that "the possibility of an operational capability in 1958 had been downgraded largely but not exclusively because of the small number of firings."[24]

Symington termed the information in Dulles's briefing "incredible." Dulles recorded Symington's analysis of the import of the new intelligence data:

> In view of his [Symington's] experience in and contacts with industry, he felt that the Soviets couldn't be making ICBMs if they weren't firing them. Considering the estimated cutback in bomber production in conjunction with our figures on ICBM firings,[25] it sounded as if Khrushchev was violating Teddy Roosevelt's principle of speaking softly but carrying a big stick, since he has virtually given the U.S. an ultimatum but at the same time is reducing his armament. . . . The DCI said that Khrushchev's statements would make one think Khrushchev would want to have us know something about his capability, rather than hide it.

Still, Symington worried that Eisenhower's budget priorities had influenced the intelligence report. He remarked wryly: "The slackening of ICBM testing and cutback in bomber production was a wonderful thing to believe if it was more important for the U.S. to balance the budget then [*sic*] to have national defense."[26] Although the raw intelligence on tests "puzzled" Dulles and Scoville, none of the intelligence officials offered even vague support for Symington's views. Major General Walsh believed the Soviets would need "20 to 50 ICBM test firings" to achieve an operational ICBM. Scoville "assured him [Lanphier] that nobody in his shop thought the Soviets had fired 55 ICBM tests."

The intelligence picture supplied by Dulles and his aides failed to convince the influential senator, however.[27] The CIA figures on Soviet missile tests drew his strongest criticism. If there had been so few ICBM tests, he observed, "it didn't make sense to estimate so much production in a short time with so little testing." Toward the end of the meeting, Symington and Lanphier emphatically restated their position. Dulles recorded that they

> emphasized that the intelligence figures on tests couldn't be right if the production estimate was right. The Senator thought intelligence was in for a surprise on the position the Soviets have today as against what we think

> they have. The Soviets were not going to give up their missile lead, but the intelligence story on testing made it look as if they were. He said he knew the U.S. was not doing enough in national defense, but implied that he couldn't get anybody to agree with him about the state of our defenses.[28]

This December 1958 meeting ended a five-month private courtship of Senator Stuart Symington by the Eisenhower administration. Symington and the CIA could not reconcile their views on the relationship between raw intelligence and estimation. The CIA argued that the date the Soviets *could* achieve a large operational ICBM force (500 missiles) was contingent on flight-testing—which could be observed and measured—but that estimates of missile capabilities in no way indicated a clear Soviet intent. Symington and Lanphier believed that the Soviets *would* deploy large numbers of ICBMs—consequently, any contradictory intelligence must be incorrect, including the new NIE. Convinced of his force projections, Symington concluded that the intelligence evidence must be wrong. His concern that the CIA estimates had been influenced by White House pressure for a smaller budget would prompt him to make the dispute public in the upcoming congressional session (see Chapter 4).

The CIA issued NIE 11-4-58 on December 23, 1958. In light of the Soviet Union's erratic flight-test program, the NIE cited "softer" evidence (the training of missile units and the "capacity" to build missiles and support facilities) as the basis for the prediction that the Soviets would have a large ICBM capability in the near future.[29] NIE 11-4-58 anticipated a Soviet deployment of 100 ICBMs by late 1959 or early 1960—a force probably consisting of SS-6s having an estimated range of 5,500 NM, a CEP of 5 NM, a postlaunch reliability of 50 percent, and a 500–800 KT warhead.[30] The SS-6 could hardly be considered a first-strike weapon against military targets; however, the NIE estimated that by 1962 the Soviets would be able to deploy 500 ICBMs with warhead yields of up to 1.5 MT, a lower CEP (3 NM), and an improved postlaunch reliability 70 percent.[31] Thus, the NIE presented a mixed evaluation: the Soviets might deploy 500 improved ICBMs but they would probably do so a year later than previously estimated. Regardless of the delay, a Soviet force of this magnitude and quality would constitute a major threat to the United States.[32]

NIE 11-4-58 included new intelligence data and estimates of Soviet missile submarine capabilities. It reduced the submarine cruise missile range estimate from 500 NM to 200 NM. Only two Soviet submarines had been converted for this purpose, but Allen Dulles and Chairman of the Joint Chiefs of Staff (CJCS) Nathan Twining thought the number could increase to between ten and twelve in the next sixteen months.[33] NIE 11-4-58 predicted Soviet deployment of a true SLBM by 1961 even though, according to Twining, "there is little direct evidence of submarine-launched missile development in the U.S.S.R."[34] American intelligence expected Soviet first-generation SLBMs to have a range of 1,000 NM, a warhead yield of 500 KT, an operational reliability of 75 percent, and a CEP of 3 NM (at ranges of 250 NM to 800 NM) or 4 NM (at ranges of 800 NM to 1,000 NM). The CIA anticipated deployment of seventeen "modern" missile submarines—six of which would be nuclear propelled—in 1960 (curiously, one year before the SLBMs would be ready, according to the NIE).[35]

Intelligence on Soviet missile programs presented a contradictory picture in 1959. On the one hand, U.S. military and intelligence had begun to sight deployed Soviet SLBMs in 1958—two years *earlier* than had been predicted in the most recent NIE. On the other hand, the United States received only sporadic intelligence on Soviet ICBMs. Soviet tests of long-range missiles (ICBMs) had stopped after the December 1958 series and did not resume until some time in the spring of 1959.[36] Intermittent U-2 flights had failed to uncover any evidence of missile base construction. Consequently, Allen Dulles occupied the unenviable position of having to reduce ICBM estimates because of erratic test patterns for the second time in two years—and just after Symington and others publicly hammered the administration with charges of covering up a missile gap.

In the summer of 1959, Dulles appointed an ad hoc panel to review a report of the Guided Missiles and Astronautics Intelligence Committee and to check the disparity between previous NIEs and the most recent intelligence data. The DCI's Ad Hoc Panel on the Status of the Soviet ICBM Program reaffirmed that "the Soviets are following an orderly, and effective ICBM program, and intend to acquire a substantial capability at the earliest reasonable date." The panel members grudgingly agreed that the Soviet Union might soon have some operational ICBMs but concluded that this "represents only a highly limited capability."[37] More important, they rejected the NIE 11-4-58 prediction of a 100 ICBM force deployment by late 1959 or early 1960. An "effective force in the international field (100 missiles) will probably not be available [to the Soviet Union] until late 1960 or later." Finally, they linked ICBM deployments to Soviet intentions (a conclusion that had been intentionally avoided in previous NIEs): "It is also believed that the Soviet determination as to their balanced needs may result in a deployment of not more than 400 to 500 ICBMs, which could be attained by the latter part of 1962. The Panel no longer believes that this latter capability will be obtained in two years after IOC, as the evidence is now firm that the Soviets are not engaged in a 'crash' program." The panel based this conclusion on two facts. First, the Soviets had tested only one ICBM design. Second, "positive evidence relative to Soviet ICBM production facilities or operational deployment sites continues to be missing."[38] Consequently, the DCI's ad hoc panel advocated further downward revision of estimates of Soviet ICBM deployments.

While the CIA struggled to comprehend the implications of Soviet ICBM test patterns, it unexpectedly received information on Soviet SLBM deployments. The military began sighting Soviet submarines with irregular features some time in 1958 and continued throughout 1959. By November 1959, the JCS had received photographs of these suspicious submarines and identified them as modified "Z" or Zulu class submarines. The Soviet Union terminated construction of this class in 1957 but, as the photographs made clear, had altered them to make them capable of launching ballistic missiles.[39] In a rush to produce an operational SLBM, the Soviets deployed a first-generation SLBM that failed to match the U.S. Polaris submarine. The Soviet SS-N-4 "Sark" carried a 2.0–3.5 MT warhead a distance of only 350 NM—not even close to the 1,000 NM. range of Polaris.[40] To achieve early deployment, the Soviets sacrificed operational efficiency and safety. The SS-N-4's liquid propulsion system reduced the length of tours at sea, required that the submarine surface to launch a missile, and increased the danger of accidents. The Soviet Union limited deployments to four submarines, two each to its Northern and Pacific fleets.[41]

The Soviet SLBM program reached other milestones in 1959. The Soviets deployed their first nuclear-powered submarine ("H" or Hotel class) as well as a new conventionally powered submarine ("G" or Golf class). American intelligence reports expected the new classes, but not the SLBM capability. Indeed, on the basis of early intelligence reports on the Golf and Hotel classes, they could not conclusively determine the missile capability of either.[42]

The negative ICBM evaluation by the ad hoc panel and the positive SLBM reports complicated ONE's task of developing the 1959 NIE on Soviet military programs. Some members of Congress and Air Force officials had challenged previous downward revisions. The combination of seemingly conflicting evidence and contentious political climate delayed the publication of NIE 11-4-59 by several months. Issued on February 7, 1960, it included a projection of the Soviet Union's intentions concerning missile deployments and stockpiles over the next four years, as in the report of the ad hoc panel. This broke from previous NIEs which estimated a date by which the Soviets would reach some arbitrary level (e.g., 100 or 500 ICBMs). NIE 11-4-59 included the ICBM and IRBM estimates for the three additional years as ranges to accommodate for future adjustments in the pace of Soviet missile activity and as a sop to placate members of the intelligence community who disagreed with the downgrading.

The ICBM, IRBM, and SLBM projections in NIE 11-4-59 did not please any branch of the armed services. The Navy, Army, State Department, and the Joint Staff of the JCS supported the lower range of missile estimates. The Air Force, still holding the mistaken belief that the Soviets had initiated a crash program, argued in favor of the higher range. But even the lower range of ICBM estimates represented a vastly inflated Soviet capability. The NIE projected a Soviet deployment of 35 ICBMs in 1960 based in part on the assumption that number of flight tests would reach 21 (and that about one-fifth of them would be failures).[43] The number of deployed Soviet ICBMs was expected to increase to between 350 and 450 in 1963. According to the NIE, Soviet ICBM forces might be smaller than previously estimated, but a smaller force could be equally dangerous because of advances in guidance, increased reliability, and a reduced CEP.[44] First-generation missile reliability was estimated to be between 50 percent and 65 percent, with increases to between 65 percent and 80 percent in three years. Estimates of CEP indicated a similar improvement: an initial decrease from 3 NM to 5 NM, and another decrease to 2 NM within three years.[45] An increase was predicted in the IRBM force; by 1963 the Soviet Union was expected to have deployed 250 such missiles and to have another 500 in inventory.[46] The report also stated that the Soviets had no crash program for developing ICBM but were pursuing a missile posture to deter the United States.[47] After attending a briefing on this NIE, presidential science adviser George Kistiakowsky made his often quoted remark: "In fact the missile gap doesn't look to be very serious (I hope this estimate is not a political effort to cut down on trouble with Congress)."[48]

NIE 11-4-59 included a projection of significant growth in "G" class submarine deployments in the next three years: 6 submarines at the end of 1959, 9 in 1960, 15 in 1961, and 18 in 1962.[49] Intelligence officials believed each submarine would carry 5 SS-N-4 missiles. Intelligence anticipated this deployment scheme would increase the number of Soviet SLBMs from 30 to 90 within three years, as well as add significant qualitative improvements.[50] Allen Dulles told Congress: "Based on our understanding of [the Soviets'] requirements and their technical capabilities, we estimate that in the 1961 to 1963 period, the U.S.S.R. will first achieve a weapons system combining a nuclear powered submarine with a 500- to 1,000-nautical-mile ballistic missile [SS-N-5 'Serb'], capable of launching from submerged positions."[51] These qualitative changes would increase the severity of the Soviet SLBM threat by compounding problems for U.S. warning and defensive operations.

The downgrading of intelligence estimates concerning Soviet ICBMs increased after the release of NIE 11-4-59. The U-2 flights gradually ruled out possible areas for ICBM base construction, but enough of such sites remained in 1960 to be an objective of Francis Gary Powers's ill-fated mission.[52] The weight of accumulating negative evidence began to take a toll on the intelligence services. Virtually all agencies agreed that the evidence dictated lower ICBM estimates, but they disagreed over how much lower. NIE 11-4-60 and NIE 11-8-60, both published on August 1, 1960, contained three separate estimates of the increase in Soviet ICBM deployments from 1960 to 1963: the Air Force predicted an increase from 35 in 1960 to 700 in 1963; the CIA predicted an increase from 30 to 400; and the State Department, Joint Staff, Army, and Navy together predicted an increase from 20 to 200. Coordination could no longer paper over disagreements about NIEs within the intelligence community. NIE 11-8-60 also estimated an increase in Soviet missile reliability during this period.[53] But these all overestimated the Soviet program. A revised NIE 11-4-60, released on December 1, 1960, included the same ICBM deployment projections but also acknowledged that such forces were primarily a deterrent and would not alter Soviet behavior.[54]

Even with four years of intelligence supplied by U-2 fights, the 1960 NIEs still contained inflated projections of Soviet ICBM strength in both the present and the future. The end of the inflation began in August 1960 with the launch of the Discoverer XIV reconnaissance satellite. In one day, it photographed 1 million square miles of the Soviet Union—more than the U-2 accomplished in all its missions, according to Dino Brugioni.[55]

In retrospect, U.S. intelligence data was either nonexistent or deficient in several key respects. Throughout the missile gap period, U.S. estimates of the number of Soviet ICBMs consistently advanced the emergence of a large force farther into the future. NIEs published from 1957 to 1960 included overestimates of the size and pace of Soviet ICBM deployments in the late 1950s and early 1960s. The errors in the NIEs can be traced to several sources, some of which were beyond the control of the CIA and the rest of the intelligence community.

Even though the U-2 flights made possible an unprecedented look at the Soviet Union, they could not satisfy the needs of U.S. intelligence. The CIA directed U-2 flights in the Soviet Union to obtain information on targets other than missiles. In classified testimony to Congress after the U-2 affair, Allen Dulles listed five targets of U-2 missions: bombers, missiles, atomic energy, submarines, and air defense.[56] Although several targets could be surveyed on a single U-2 flight, the small number of penetrations (about thirty, according to Lawrence Freedman) meant that much Soviet activity remained hidden from the aircraft.[57] Further, a U-2 plane could effectively take photographs only about five hours per day. Seasonal conditions in the Soviet Union also limited the times when missions might be conducted.[58] Allen Dulles admitted the limitations of U-2 flights in testimony before a closed session of Congress: "So we have clues to what we want to know, but we have not even completed the photography of all of the Soviet Union we would like to photograph. . . . There is a lot we don't know."[59] For these reasons, the intelligence data supplied by U-2 flights could not disconfirm projections of Soviet ICBM deployments. Raymond Garthoff, an intelligence officer involved with the NIEs, writes: "But we could not persuade others in the estimates process, nor could the president persuade skeptics, because the evidence was indicative and substantial, but not conclusive. Available evidence was not sufficient to rule out other possibilities. Although those who advanced more dire estimates of existing (and especially future) Soviet missile capabilities could not, of course, prove their case, it could not then be completely disproved."[60]

Intelligence officials committed several errors in evaluating the relationship between Soviet missile quality and missile deployments. In their early analyses, they claimed that the Soviets had chosen ICBMs of simple design with a high degree of reliability. Dulles informed Eisenhower in October 1958 that "the Soviets have made maximum use of proven components and of their considerable previous experience with shorter range missiles." As a result, "the USSR will not necessarily require a large number of ICBM flight tests before entering into production."[61] The problem, arose when the Soviets continued testing the first-generation SS-6 ICBM rather than deploying it. Although simple in design, the SS-6 was neither easy to deploy nor very reliable. When SS-6 deployments failed to materialize as expected, Allen Dulles revised the interpretations of Soviet objectives.[62] Apparently, SS-6 design problems were not reflected in CIA estimates until 1960. But the CIA never concluded that lack of operational reliability had prevented SS-6 series production and deployments. The absence of deployed missiles led the CIA to anticipate higher quality once the missiles were deployed.

In addition, the estimates of ICBM deployments endured competing intelligence bureaucracies, each with a particular perspective and interpretation. This problem became especially acute in 1959 and 1960. The weak data base, coupled with the absence of Soviet ICBM deployments, encouraged different segments of the intelligence community to promote alternative estimates. The coordination of such estimates allowed influential actors to press their case and to seek bureaucratic allies. In the process, disagreements solidified and became more permanent. But one should remember that *all* intelligence projections of Soviet ICBM deployments from 1957 to 1960 were wrong, although some were much more wrong than others.[63]

An examination of the entire range of Soviet missile programs (ICBM, SLBM, IRBM) reveals that systematic bias within the intelligence community cannot account for the inflated ICBM estimates. While the intelligence community consistently overestimated Soviet ICBMs, it predicted accurately—and sometimes even *underestimated*—Soviet SLBM developments. For example, in NIE 11-4-59, it underestimated the scale of Soviet ballistic missile submarine deployments. From 1960 to 1962, the Soviet force (not including the "Z" class) increased from 10 submarines to 24, not from 9 submarines to 18 as predicted in the NIE. The number of Soviet SLBMs in 1962 was only marginally larger than predicted in NIE 11-4-59 (72 SLBMs rather than 60), since "G" class submarines carried 3 SLBMs—not 5 as in the intelligence estimate.[64] But whereas a deployment of 250 IRBMs in both 1962 and 1963 (with inventories of 690 and 750, respectively) was predicted in NIE 11-4-59, the Soviets deployed 472 IRBMs in 1962 and 682 in 1963.

The inaccuracy of NIEs can also be attributed in part to the strategic choices made by the Soviets beginning in the summer of 1958. They apparently decided to emphasize intermediate- and medium-range ballistic missiles, probably because of technical problems in the ICBM program—as well as for geopolitical reasons. The Soviets probably calculated greater political benefits in the near term from IRBMs targeted against Europe.[65]

As intelligence officials advanced the achievement of a large Soviet ICBM force farther into the future, the probability of numerical ICBM superiority in favor of the Soviet Union decreased because of the success of U.S. programs for securing strategic retaliatory forces (by accelerating the production of first-generation ICBMs and Polaris submarines, developing the second-generation Minuteman ICBM, dispersing SAC bombers, and instituting a ground alert system). However, the decreasing probability of a numerical missile gap in favor of the Soviet Union did not lessen concerns about U.S. strategic vulnerability.

Warning in the Missile Age

To appreciate fully the perceived threat from Soviet missiles, we must turn our attention to the capacity of the United States to detect a Soviet attack. Without adequate warning, the United States might not successfully deter the Soviet Union because it would not be able to assure a significantly large retaliatory strike. Warning was critical to ensure retaliation because the Eisenhower administration had little hope of having an effective defensive system before the anticipated Soviet missile deployment.[66] During the missile gap period, the administration expended large sums of money and considerable effort to improve systems for tactical (immediate) and strategic (long-term) warning of Soviet bomber and missile attacks. Both warning systems were given a higher priority immediately after the launch of Sputnik.[67] Despite these efforts, the systems suffered from numerous deficiencies, in large part because of the complexity of the potential Soviet threat.

The U.S. programs for tactical warning in the pre-Sputnik period concentrated on long-range bombers—the only means of delivering nuclear weapons then available to the Soviet Union. In the 1950s, the United States and Canada built three different radar systems to warn of a bomber attack: the Pinetree Line, the Mid-Canada Line, and the Distant Early Warning Line. These systems seemed adequate when the bomber gap failed to materialize and the Soviet bomber force remained under 200 aircraft. But weaknesses in the systems' design and performance caused much consternation among Washington policy makers, who anticipated that Soviet bombers might exploit gaps in the areas and altitudes covered by the three radar networks and thus reduce the warning of an attack. A panel of the President's Science Advisory Committee (PSAC) determined that "low-level radar coverage is essentially worthless."[68] In a hypothetical Soviet bomber attack staged simultaneously at low altitudes (2,000 feet) and high altitudes (50,000 feet), the warning times for the four B-52 bases servicing the nuclear war plan were: Westover, 54 minutes; Loring, 26 minutes; Fairchild, 108 minutes; Castle, 29 minutes.[69] In such a case, only a handful of B-52s would escape destruction and be able to retaliate. Other measures could also be employed to reduce warning of a Soviet bomber attack. A series of SAC bomber exercises revealed that electronic countermeasures "virtually blinded the defensive radars."[70] NORAD tests conducted in the fall of 1958 showed that the coordination of information during a bomber attack would be very difficult. In one exercise, NORAD's tracking system became overloaded. After reviewing the situation, the members of a PSAC panel reported that "we should realize that tactical warning of an attack [by bombers] is by no means assured. We are convinced that in the absence of a greatly increased level of air defense a well-planned sneak attack by the Soviet air force, preceding a planned mass attack, might well escape detection and result in the destruction of a large portion of SAC aircraft on the ground."[71] The United States never rectified the problems affecting bomber warning systems because ICBM warning systems received a higher priority.

The nation's policy makers knew that early warning would be an essential element of deterrence in the missile age.[72] The short flight time of ICBMs (about thirty minutes) imposed new constraints, which affected decision making and the disposition of nuclear forces. Each extra minute of warning time was valued for its potential to ensure that there would be a retaliatory strike and that it would be of sufficient magnitude.

Several months after the launch of Sputnik, the Air Force proposed the construction of three Ballistic Missile Early Warning System (BMEWS) radars as part of its "Missile Defense System."[73] The radars would be located at Thule (Greenland), Clear (Alaska), and Fylingdales (England). All three would track ICBMs fired over the North Pole; the last would also track IRBMs aimed at Western Europe. When the estimated costs of the program skyrocketed from $533.7 million to $1.328 billion within six months, the Defense Department warned the Air Force that the program would be canceled unless the costs were reduced. The Air Force and the Defense Department decided to eliminate the third radar at Fylingdales in order to bring the program costs down to $822.7 million.[74] But shortly thereafter, scientific analysis determined that complete coverage of the United States could be attained only with activation of the third BMEWS radar; ICBMs fired in lower trajectories and aimed at certain parts of the United States would not be detected without it. According to the Weapons Systems Evaluation Group (WSEG), a critical situation would arise once the Soviets had deployed a substantial ICBM force: "Without this third BMEWS site, a serious gap in coverage will exist. For example, calculations show that 5,500-nautical mile, 23 degree trajectory, ballistic missiles could be launched from Western Russia to escape detection entirely by the planned Clear and Thule BMEWS sites so as to impact on 47 SAC bases without warning (only 15 western U.S. SAC bases are fully protected by these sites). Against low angle (15 degrees) trajectories, even fewer bases are fully protected."[75] When it was shown this analysis, the DOD allowed the Air Force to reinstitute the plans for the third radar.

Because of the pressing need for ICBM warning systems, BMEWS received high priority and additional funding. The Thule and Alaska sites became operational as scheduled (on December 31, 1960, and June 1961, respectively). However, the temporary cancellation and negotiations between the United States and Great Britain over construction of the third radar and division of operational responsibilities pushed its projected initial operating date into 1963. (It became operational on September 15, 1963.)[76] The PSAC subsequently found weaknesses in the BMEWS and recommended additional measures (Midas warning satellites and a U-2 infrared warning aircraft) to compensate for them.[77]

The development of measures to counter the Soviet SLBM program posed an entirely different problem for the United States. Submarine mobility and the short flight times of SLBMs meant that the United

States could not rely on missile warning alone. Antisubmarine warfare became as important as SLBM warning systems in the attempt to counter Soviet ballistic missile submarines.[78]

The U.S. attempt to locate Soviet submarines achieved some success in 1957 with the deployment of the Sound Surveillance System (sosus)—a network of hydrophones located at strategic positions on the ocean floor that can identify Soviet submarines by the noises they emit. Research on this system began in the early 1950s and received the endorsement of the Killian and Gaither panels.[79] But the deployment of sosus was not sufficient to bolster policy makers' confidence. WSEG-30, according to the WSEG director, reported that "With only the air cover presently employed on a routine basis to backup sosus, the probability of our timely detection of a submarine force of perhaps 20 submarines while entering our coastal waters is low."[80] In early 1959, a PSAC study group, "Project Atlantis," concluded that "ocean area submarine surveillance is now technically feasible, requiring new and major effort for its installation."[81]

Effective ASW was certainly not an immediate possibility for the United States, largely because of organizational and technological deficiencies. General Thomas Power, the SAC commander, complained in 1957 that the weaknesses of ASW and a lack of warning compromised half of SAC's bases. He also told Air Force Chief of Staff Thomas White: "It appears that the protection of our forward bases from this threat is being subordinated to an effort to maintain a posture for the application of naval forces to an offensive role."[82] But increased cooperation between the armed services could not compensate for technological deficiencies. As CJCS Twining informed Defense Secretary McElroy, "Various studies and reports in progress confirm that for a major advance in antisubmarine warfare there is a need for a major scientific breakthrough to enable long range wide area underwater surveillance, both fixed and mobile, as a prerequisite for initial detection and subsequent localization of deeply submerged silent running submarines."[83]

Detection by means of both ASW and sosus became more difficult after 1959 because the Soviets had developed nuclear power—which meant quieter and longer submarine operations—as well as missiles with greater range, thus complicating the SLBM warning problem. On March 20, 1959, the WSEG, in Report no. 35 ("Defense against Sea-Launched Missile Attack"), concluded that some tactical warning might be achieved by deploying the U-2 Infra-Red system proposed by the PSAC for ICBM warning. But CJCS Twining questioned its value, even though the United States did not then have an SLBM warning system in place.

> Such a system, to be effective, should provide sufficient warning time so that a substantial number of strike aircraft can take off and clear the prospective blast area before the missile reaches its target. WSEG-35 analyzes this problem and demonstrates that unless five minutes warning at an airbase is available, little benefit accrues.
>
> Ballistic missiles travel approximately one hundred miles per minute. Using WSEG-35 criterion, little benefit would be derived from instantaneous warning of a firing by a submarine within 500 miles of its target. The minimum acceptable warning time would be five minutes plus the time necessary for: (1) evaluation by the detecting station on the firing indication, (2) transmission of the warning, (3) evaluation by the receiving station, and (4) decision and issuance of the alert. For each minute of time consumed in these four actions, an additional 100 miles must be added to the basic 500 miles within which warning would be of little benefit.[84]

The problem of constructing an SLBM warning system seemed insurmountable. Not surprisingly, the United States did not deploy such a system until 1970.[85] With problems in detection, ASW, and warning, few viable options were available to lessen the potential effect of the Soviet SLBM threat.

A system that would ensure tactical warning of a Soviet missile or bomber attack remained essentially nonexistent throughout most of the missile gap period. The Eisenhower administration recognized that tactical warning uncertainties made strategic warning "increasingly important." According to NSC 5802/1, "even if some risks have to be taken, vigorous efforts should be made . . . to collect and accurately evaluate indications of hostile intentions that would give maximum prior strategic warning of hostile action against the United States."[86] Several steps were taken to improve strategic warning. Eisenhower agreed to a series of meetings with representatives of the Soviet Union in Geneva to discuss the subject of surprise attacks in the summer of 1958. The scope and ambiguity of their objectives prevented the achievement of even marginal success, however; the talks collapsed the following January.[87] The National Indicators Center, which was established in 1954 to monitor Soviet actions that might give strategic warning, encountered a number of organizational and structural difficulties.

Its reliance on other intelligence agencies for information meant that it could not "assure prompt transmission of needed information."[88] The insensitivity to potential indicators became clear when three of twenty-four trans-Atlantic telephone cables were cut within three days in 1959; the intelligence community, including the National Indicators Center, did not know about these actions for several days. The United States had some capacity for strategic warning, but it would be erroneous to conclude that the Eisenhower administration simply fell back on strategic warning when tactical warning became problematic.

The Perceived Vulnerability of United States Strategic Forces

When analysts and policy makers combined the inflated NIE projections with warning deficiencies and applied the result to the nuclear force posture of the United States, they produced worst-case scenarios of the Soviet threat. Variations in any one of these factors—each of which was dependent on other variables—could substantially affect perceived vulnerability. Policy makers monitored them closely to detect any changes, especially since the nation's strategic nuclear force was composed almost entirely of vulnerable SAC bombers. A "critical period" was never reached, largely because of the U.S. and Soviet policy choices and programs. The worst-case scenarios are important nevertheless because they help to explain how the perception of vulnerability impacted on policy makers and why it persisted throughout the missile gap period.

Eisenhower administration officials scrambled to evaluate the nation's strategic vulnerability in the aftermath of Sputnik. Robert Cutler, Eisenhower's special assistant for national security affairs, reported to the president that 1654 bombers (214 B-52s, 180 B-36s, and 1,260 B-47s) were located at SAC bases in the United States—or the zone of interior. Only four bases (Castle, Westover, Loring, and Fairchild) served the 119 B-52s prepared to execute the existing war plan.[89] Under conditions of attack, SAC could release 134 loaded bombers (8 B-52s, 9 B-36s, and 117 B-47s), but only *if* bases overseas and in the continental United States received warning times of a half hour, and two hours, respectively. CINCSAC Thomas Power confirmed Cutler's calculations. An SAC alert test conducted on October 1, 1957, indicated that 129 bombers carrying 138 weapons would be able to retaliate against the Soviet

Union. Since many of these aircraft were located at vulnerable overseas bases, Power worried that restrictions on the employment of nuclear weapons would prevent the launch of some of them. He warned that JCS restrictions on nuclear weapons "will do little but make scapegoats of the operational commanders in the event of a modern Pearl Harbor."[90] Cutler estimated that the entire U.S. strategic deterrent represented only 56 targets for the Soviet Union.[91] The implications were clear: it would require only a handful of Soviet nuclear weapons to neutralize the backbone of the U.S. strategic deterrent. Upon receiving Cutler's information, the president demanded that the Air Force submit a report on the status of overseas bases by November 4, 1957, at 9:30 A.M.—one half hour before his meeting with the Science Advisory Committee of the Gaither panel.[92]

The Gaither panel emphasized SAC's vulnerability and endorsed a series of measures aimed at reducing it. The measures designed to strengthen SAC centered on increasing its offensive capabilities and making SAC bases less vulnerable to attack by Soviet bombers and missiles. The report called for the expansion of radars to guarantee tactical warning and Nike-Hercules or Talos missiles for the defense of SAC bases. Before a Soviet ICBM threat materialized as anticipated in 1959, the United States should institute a radar system to ensure early warning of an ICBM attack, develop a bomber alert program on the order of 7 to 22 minutes, disperse aircraft facilities with concrete shelters hardened to the range of 100 to 200 pounds per square inch, and establish an active missile defense against ICBMs for all bases.[93]

Gaither panel members Robert Sprague and William Foster withheld the most critical information about SAC vulnerability from the report. After the panel had made its presentation to the NSC, Sprague and Foster personally delivered "a special highly sensitive report" to the president concerning the ability of SAC to respond to a Soviet first strike. They told him that during periods of low international tension, SAC might be unable to deliver a retaliatory blow against the Soviets if it received too little warning. Sprague noted that in September, SAC had required six hours to get any planes airborne. Further, the proximity of overseas bases to the Soviet Union made them unreliable for U.S. retaliatory missions. He estimated that the Soviets could easily devote four nuclear weapons to each of SAC's sixty bases. In these circumstances, he concluded, the United States might be able to retaliate with 50 to 150 nuclear weapons, but "we cannot assume we could lay down a substantial retaliatory attack."[94] After the two left, Secretary of State John Foster Dulles told the president that he considered the probability of a full-scale Soviet attack during a period of low international tension "so remote" that the cost of the alert concept was unjustified.[95]

Two WSEG reports issued in 1958 contain characteristic evaluations of the Soviet missile threat after Sputnik. In August, the WSEG estimated that the SAC could launch 350–400 bombers after receiving tactical warning of Soviet attack. Still, the United States might lack a credible retaliatory capability as soon as sixteen months thereafter, possibly lasting for two years:

> During 1960, when the Soviets' estimated potential capability for building ICBM's might allow them to have about 100 missiles, our manned bomber force may be especially vulnerable to surprise attack because the BMEWS system will not be completely operational before 1961. . . . It is thus seen that a critical period may exist some time between 1960 and 1962 when the Soviets may possess a significant ICBM capability, [*sic*] our own ICBM capability (under presently approved DoD programs) is limited and our BMEWS may not be in operation. Whether or not this critical period develop's will depend upon the Soviets' actually achieving a high quality ICBM system and deploying missiles in numbers for which they may have the potential.[96]

In October, the WSEG confirmed this analysis, albeit in a less alarmist tone. It referred to the "extreme importance of warning": "Assuming no strategic or tactical warning, 100 ICBM's might severely reduce the SAC retaliatory strike force although a credible warning of 15 to 30 minutes would allow launching of a substantial SAC force. Obviously, unless an effective ballistic missile warning system is present, and SAC is in an alert posture so that it can take advantage of it, the situation would be worse with a Soviet force of 300 to 500 ICBM's possibly facing us by 1961."[97] In 1958, the critical period remained several years away. It would be reached only when Soviet missiles numbered more than several hundred, as long as U.S. retaliatory forces remained small and warning systems inadequate. Smaller Soviet deployments or expanded U.S. strategic programs would prevent the vulnerability described by the WSEG.

Changes in the intelligence estimates in late 1958 yielded new projections of the nation's vulnerability in 1959. Two reports issued in that year, a PSAC report on passive defense and a WSEG study on the strategic benefit of an additional runway at SAC bases, illustrate the evolution of the estimate of the Soviet missile threat from a force constituted solely of ICBMs to a more complex and qualitative threat. Presidential science adviser James Killian and an aide, Brockway McMillan, delivered the PSAC report to President Eisenhower on March 17, 1959. In their briefing, they evaluated the potential strategic threat facing the United States up to 1962. The United States, McMillan told Eisenhower, offered three types of targets for a Soviet strike: SAC bomber bases, air defense sites, and missile bases. The total number of U.S. targets would increase from only 40 in 1959 to 140 several years thereafter as the United States deployed ICBMs and dispersed SAC bombers. McMillan estimated that because of the low reliability and inaccuracy of first-generation Soviet ICBMs, the Soviets would need 5 ICBMs per target (or a total of 200) to launch a successful strike in 1959. This estimate far exceeded even the most dire predictions of Soviet ICBM deployments. Whereas the number of Soviet missiles necessary for an attack would increase as the United States deployed additional targets, he warned against assuming "that the Soviet force will always be inadequate for a decisive attack against our continental forces."[98] The Soviets might deploy ICBMs faster than the United States was able to expand and protect its strategic forces.

"The true strategic position [of the United States] is unclear," McMillan told the president, because of possible variations in Soviet deployments and actual warning time. But he presented a harrowing picture of the SAC force under a Soviet ICBM attack:

> When the real attack comes, every minute of indecision costs us about 40 bombing aircraft; five minutes of indecision costs us 200 bombing aircraft; every SAC base covered with snow costs us about one aircraft per minute. A complete failure of BMEWS—perhaps because of jamming or spoofing—would cost us about 450 bombers. . . . The force that escapes, therefore, is not exactly of our choosing, but is selected by the fortunes of war. It may not therefore conform in its structure to a pre-determined war plan; it may have to be regrouped and re-directed to targets for which it is inadequate, according to information and a plan developed during the attack.[99]

After the briefing, President Eisenhower expressed skepticism about attempts to estimate Soviet strategic capabilities across a five-year period because such analyses often failed to acknowledge potential problems in weapons development. But he admitted that "if we really got into a war we should get off our striking power as quickly as possible."[100] Eisenhower and his advisers agreed on the need for increased ICBM hardening and accelerated Polaris construction as insurance against future Soviet deployments.

In Staff Study no. 77, the WSEG estimated the number of additional bombers that would be released in 1962 after three types of Soviet missile attack (ICBM, SLBM, and a combined attack) against SAC bases in the zone of interior to determine the strategic benefit of a second runway at those bases.[101] A second runway would provide relief from a Soviet missile attack only if it could reduce the time between launches of alert aircraft after SAC received warning.

Using projections from NIE 11-4-58, the WSEG predicted the results of the three types of missile attack. In a Soviet ICBM attack, most ground alert heavy aircraft (95–100 percent) and a slightly smaller percentage of medium aircraft (58–97 percent, depending on missile trajectories) would escape destruction (assuming three operational BMEWS sites) even if they received only tactical warning (15 minutes). SAC's survival in the event of an SLBM attack would depend greatly on warning time. Unless a warning system was established "before the Soviet sub-launched missile threat reache[d] dangerous proportions, SAC bases could be attacked essentially without tactical warning." Without strategic warning (12 hours), an SLBM attack would constitute "an extremely grave threat to the survival of the ground alert force." Specifically: "Only 35 to 65 percent of the ground alert heavy force and 28 to 45 percent of the ground alert medium force, survive for launching."[102] Since the ground alert force constituted only one-third of SAC, only a small percentage of the entire strategic force could actually retaliate in these scenarios.

In the event of a Soviet combined ICBM/SLBM attack when only two BMEWS were operational, 47 of the 62 SAC air bases in the zone of interior would receive no warning. (A third BMEWS would provide 21 to 30 minutes' warning for these bases.) To exploit this weakness, the Soviets could target ICBMs against these 47 bases and SLBMs against the remaining 15. The maximum tactical warning for these 15 bases, all located in the western United States, would be 10 minutes, even with an SLBM warning system in place. The WSEG concluded that in this worst-case scenario, "only 18 to 32 percent of the total alert force [132 to 248 bombers] could be expected to survive."[103]

According to WSEG calculations, the outlook would be bleak for SAC if Soviet deployments were consistent with intelligence projections. Unless the United States received strategic warning, the largest bomber force to retaliate in any of the three scenarios would number only 379 (205 B-52s and 174 B-47s). In the most disadvantageous circumstances, only 71 bombers would retaliate. In any case, only a fraction of the SAC total (1,710 bombers) would be able to retaliate after receiving tactical warning. Little comfort could be gained from the WSEG's exclusion of U.S. ICBM and SLBM forces since the bomber force would still be by far the largest, most accurate, and destructive leg of the strategic nuclear triad in 1962. Even though U-2 flights and other intelligence sources indicated that no Soviet ICBMs were operational at that time, it was not clear whether American BMEWS or Soviet ICBMs would be operational first. An early Soviet SLBM IOC would bode ill for the United States. If the Soviet Union deployed as few as 100 ICBMs before late 1962, then that force, in combination with the SLBMs, would give it a capability to strike SAC bases with little or no tactical warning. The United States would have to rely on receiving strategic warning to guarantee retaliation in this situation.

By 1959, the vulnerability of the United States to a Soviet first strike had become a race between Soviet missile deployments and U.S. force expansion and BMEWS construction. NIE 11-4-59, published on February 9, 1960, indicated the extent of this potential vulnerability. The American target system for a Soviet attack was expected to increase from 72 in 1960 to 212 in 1963, but Soviet deployment of ICBMs would increase at a higher rate. According to the NIE, in 1961 the Soviets would be able to destroy 55 percent to 85 percent of the total SAC force if they deployed 140 ICBMs and 70 percent to 95 percent of the force if they deployed 200 ICBMs. Changes in the predicted level of Soviet ICBM quality accounted for the range of 25 percent in the higher damage estimates. Thus, overestimates of Soviet ICBM deployment capabilities and high qualitative estimates produced an ominous picture of the threats to SAC survival even as late as the early 1960s.[104]

Herbert York, the Defense Department's director of research and engineering, applied his own calculations to the NIE information and found that U.S. retaliatory power could be disarmed by the Soviets in 1961. York recalls: "My calculations showed that for a period beginning early in 1961 and lasting for many months the Soviets could hypothetically reduce our retaliatory forces to zero in a surprise attack. I didn't call it that, but the situation was identical to the one that in later years would be referred to as the 'window of vulnerability.'"[105] He circulated his results to other administration members in early March 1960, offering several changes in the nation's strategic posture to rectify the potential problem. After a four-hour meeting with York, presidential science adviser Kistiakowsky recorded that York "said that the studies in his office indicate that if one is a pessimist and allows the Soviets the maximum capability that the national intelligence estimates suggest, in mid-1961 we will pass through a critical period when they may be able to destroy virtually all our retaliatory forces by a perfectly coordinated surprise attack."[106] To limit SAC vulnerability, York recommended increasing Polaris deployments to five submarines by the end of 1961 and moving up the operational date of the second BMEWS site (Clear, Alaska) from September 1961 to the beginning of that year. President Eisenhower endorsed these actions.

The United States continued to be concerned about SAC vulnerability to Soviet missiles throughout the rest of 1960, even though NIEs in that period predicted a reduction in the Soviet ICBM threat. The NSC again called for an increase in ASW, given the lack of active defenses.[107] In December 1960 the WSEG, in a major study on offensive weapons systems, reiterated the seriousness of the Soviet SLBM threat that would exist in the mid-1960s. Because of the absence of SLBM warning, "the requirement for air alert in the 1964–1967 time period . . . depends critically on the magnitude of the Soviet SLBM threat."[108] A massive interdepartmental study, organized at the request of President Eisenhower and coordinated by the NSC, affirmed that a critical period might arise in the next two years:

> The principal danger now facing the U.S. arises primarily from the margin of the Soviet lead in ICBM capabilities which is expected during the 1961–1962 time period. . . . This threat to survival would be greater if substantial slippage occurs in U.S. BMEWS and offensive missiles programs, if the median intelligence estimate of Soviet ICBM capability is exceeded, and if the initial Soviet ICBM attack were executed without warning. The U.S.S.R. during 1961–1962 could execute a ballistic missile attack and a follow-up missile-bomber attack which, if undertaken without warning, could destroy or neutralize a large segment of U.S.-Allied nuclear striking forces in the U.S. and overseas, and which would bring into question the survival of the United States.[109]

These conditions were less likely to arise than those in the earlier WSEG scenarios. The interdepartmental study was never formally presented to the NSC or to President Eisenhower because of its late completion. The study would serve as "a legacy for the next administration in its consideration of national security problems of the future."[110]

In another report, published only two weeks before Eisenhower left office, the WSEG downgraded the possibility of a Soviet first strike that could cripple U.S. retaliatory forces. Three conditions would have to exist, it claimed, for a surprise attack to be "attractive" to the Soviets: Soviet strategic forces whose quantity and quality exceeded those predicted in NIE 11-8-60; no U.S. aircraft on airborne alert; and no tactical or strategic warning for the United States. The WSEG observed: "If Soviet leaders have been considering such an attack, it appears that they might have had their best opportunity for a successful surprise attack in the latter part of 1960." It also reported that trends were rapidly moving against achievement of a surprise attack capability that could avoid "unacceptable damage" in retaliation.[111]

In sum, U.S. officials perceived that SAC bases would be vulnerable to a combined ICBM/SLBM attack by the Soviet Union in the near future *if* it deployed a minimal number of ICBMs *and* the United States failed to take preventive measures to provide warning and protect its retaliatory forces. President Eisenhower, who displayed a healthy skepticism of intelligence projections during the entire missile gap period, was persuaded of the import of a potential strategic imbalance and responded cautiously. The Eisenhower administration instituted changes in warning systems and offensive deployments to ensure that Soviet deployments would not undermine the forces' retaliatory capacity.

Inflated NIEs, a weak and uncertain warning system, and the U.S. strategic posture combined to produce worst-case projections concerning U.S. vulnerability. These projections changed with changes in the Soviet ICBM program. But in such projections, Soviet SLBM deployments and high estimates of ICBM quality compensated for slow ICBM deployments. As a result, the primary concern moved from a singular ICBM threat to a combined ICBM/SLBM attack; moreover, it was assumed that qualitative improvements in Soviet weapons performance would decrease the number of ICBMs needed for a Soviet first strike. Early intelligence assessments projected such poor performance by ICBMs that, according to them, the Soviets could gain a military advantage only if they possessed large numerical superiority and struck first. The NIEs increasingly attributed more dangerous capabilities to the first-generation missiles, even though the Soviet Union had not yet deployed any ICBMs. The qualitative improvement in Soviet ICBMs predicted by the NIEs changed the assessments of U.S. strategic vulnerability. Improvements in ICBM performance would increase the military effectiveness of initial deployments; thus smaller numbers of such missiles could threaten SAC's retaliatory capacity. But delays in ICBM development by the Soviet Union and acceleration of strategic programs by the United States essentially precluded this outcome.

Another reason for the strategic vulnerability pointed to in these projections, particularly those in WSEG Staff Study no. 77, was the anticipated Soviet SLBM deployment, which complicated the strategic problems confronting the United States. Because of their short flight times, SLBMs could pose a grave threat to the retaliatory capacity of SAC.

Perceptions of U.S. vulnerability during the missile gap period were more complex and uncertain than most scholars have recognized or been willing to acknowledge. The conventional wisdom has emphasized bureaucratic causes for having produced inflated intelligence estimates and has often accepted U-2 intelligence as decisive. It has largely ignored weaknesses in intelligence information and the problems of U.S. warning systems—both of which are important for understanding why decision makers' concerns about strategic vulnerability persisted in these years.

Recently declassified documents provide a much more complex picture of the nation's strategic vulnerability in the late 1950s. The worst-case scenarios were the product of three circumstances: inflated intelligence estimates of Soviet missile deployment, inadequate attack warning systems, and the U.S. strategic force posture. Vulnerability projections would not have been so dire had the status of programs been different or the intelligence process been more accurate. But uncertainty about Soviet missile progress, the construction of tactical warning systems, and the diversification and protection of U.S. strategic forces prompted the Eisenhower administration to consider vulnerability estimates very seriously. The legacy of Pearl Harbor and Cold War enmity meant that worst-case scenarios could not be dismissed so long as outcomes remained in doubt.

The relationship between vulnerability and uncertainty is clearest in the NIEs on Soviet missile programs. American intelligence estimates suffered from a series of problems which, cumulatively produced wildly inflated projections of Soviet ICBM capabilities. A weak and incomplete intelligence data base, erroneous assumptions made in analyzing technical progress, and bureaucratic competition in the coordination of estimates resulted in NIEs that, throughout this period, overestimated virtually every aspect of Soviet ICBMs. Intelligence estimators continually revised their projections of Soviet ICBMs to account for new information, but such projections were never accurate. Yet estimators accurately projected—and at times even underestimated—Soviet SLBM deployments. While the Soviet ICBM program remained enigmatic, the Soviets deployed new submarine models and SLBMs from 1958 to 1961—much to the surprise of the United States. Indeed, U.S. intelligence officials struggled to determine the number of SLBMs carried by the new subs. The ICBM and SLBM projections reflect the difficulty and uncertainty of estimating military capabilities in a closed society in the pre-satellite era.

Compounding this uncertainty was the problem of making assumptions about Soviet intentions. The ultimate objectives of the Soviet Union and Khrushchev were implicit considerations in all the NIEs. Before 1960, intelligence officials tried to avoid the issue, claiming that they simply projected maximum deployments based on full production. But assumptions about Soviet intentions crept into the estimates—in part because of Khrushchev's effective disinformation campaign. Intelligence officials operated on the assumption that the Soviets intended to deploy a large missile force as soon as was practical, though they might not institute a crash program. This led to projections of deployments several years in the future, well beyond the period covered by available intelligence data except, perhaps, for some guesses of Soviet missile production capacity.

If Soviet missile capabilities and intentions mystified intelligence officials, the status of U.S. attack warning systems worried them very much. They had little confidence that a Soviet attack would be detected during the missile gap period. The bomber warning radar systems (Pinetree, Mid-Canada, and Distant Early Warning lines), which were fully operational before the launch of Sputnik, could be fooled or circumvented by both active and passive means. The United States did not achieve partial ICBM warning until December 31, 1960—only six weeks before Defense Secretary Robert McNamara declared the missile gap an illusion—because development of the BMEWS started *after* Sputnik. Programs for countering Soviet SLBMs (ASW, SOSUS, and a warning system) progressed at an even slower pace during this period.

Consequently, warning systems created additional uncertainties for the Eisenhower administration. Would the BMEWS become operational before the Soviets deployed 100 or more ICBMs? How many ICBMs could the Soviets deploy before the BMEWS provided complete coverage of the United States? Would Soviet SLBM development and deployment overwhelm ASW and SOSUS? Could high- and low-altitude gaps in bomber warning lines, as well as other problems, be corrected quickly? From 1957 to 1960, the Eisenhower administration could not be absolutely confident that the answers to these questions would be favorable.

Uncertainty about the intelligence data, Soviet intentions, and the amount of strategic and tactical warning meant that analysts could not completely disregard surprise attack as a Soviet option in the future. Changes in the Soviet strategic force posture, both real and imagined, allowed vulnerability perceptions to persist even after the Soviet ICBM threat began to decline. Worst-case scenarios based on Soviet SLBM deployments and intelligence estimates of Soviet ICBM quality preoccupied the Eisenhower administration almost to its final days.

[3]

The Challenge to Massive Retaliation

This chapter examines the Eisenhower administration's formulation of national and nuclear strategy during the post-Sputnik years. As part of the New Look, massive retaliation dominated both areas from 1953 to 1957. But Sputnik's launch changed the external environment dramatically and put massive retaliation on precarious ground. The prospective Soviet ICBM capability prompted a reassessment of whether the United States could continue to rely on the threat of nuclear escalation for immediate and extended deterrence of the Soviet Union. Would the administration's reliance on massive retaliation still be valid after the deployment of Soviet ICBMs? Would the threat still be credible in the eyes of Western Europe and the Soviet Union? If the national strategy was deficient, what alternatives could replace it? What nuclear targeting strategy could best support deterrence? And what forces would be required to implement this strategy? Finding answers to these difficult questions constituted an important test for the Eisenhower administration. Its decisions about the direction of national security policy and nuclear strategy constitute a critical stage the evolution of U.S. nuclear policy.

Massive retaliation as a national strategy is the subject of the chapter's first section. The National Security Council's policy papers on "Basic National Security Policy" (NSC 5810/1 and NSC 5906/1) and "U.S. Policy in the Event of War" (NSC 5904/1) reflect the administration's efforts to analyze nuclear weapons policy in the broader context of U.S. foreign policy. These papers directly addressed the questions surrounding the administration's reliance on massive retaliation to achieve foreign policy objectives. The deliberations on these NSC papers reveal the administration's division on the question of whether a national strategy of massive retaliation lacked credibility and needed to be changed. Despite the growing concern, the Eisenhower administration ultimately rejected the challenges to massive retaliation—in effect, staying the course of the "New Look".

President Eisenhower's willingness to retain massive retaliation was based on both fiscal and strategic reasoning. Abandoning massive retaliation would have required an expansion of conventional forces, resulting in increased defense expenditures. The president also believed that any indication of reticence to employ nuclear weapons would embolden the Soviets and weaken strategic deterrence. This conviction, and his belief in the escalatory nature of modern warfare, led him to reject any de-emphasis of massive retaliation as a national strategy.

The chapter's second section concerns the high-level formulation of a nuclear strategy to support the national strategy. Two main questions faced the administration. First, what type of targeting doctrine should the United States employ against the Soviet Union? This question became especially important in the late 1950s and early 1960s, when the U.S. and Soviet strategic force postures underwent major changes. Second, what is the optimal U.S. stategic force posture, given the strength of Soviet forces and U.S. targeting plans? Or more succinctly, how much is enough? In an effort to formulate an acceptable nuclear strategy, President Eisenhower and senior administration officials examined reports on strategic force posture by the WSEG and damage estimates by the Net Evaluation Subcommittee (NESC), and considered the creation of a Single Integrated Operational Plan (SIOP). Many key concepts of modern nuclear strategy can be traced to this effort, among them the justification for a triad; the futility of attempting to achieve damage limitation by employing counterforce or active defense; the diminishing returns of offensive force procurement; and the beginning of what was to be the policy of mutual assured destruction (MAD). Ultimately, the president became increasingly frustrated by a nuclear strategy that produced war plans with enormous overkill.

President Eisenhower's inability to mold nuclear strategy to conform to his desires can be attributed to the rivalry between the armed services and the ambiguity of the doctrine of massive retaliation. Nuclear strategy, as the armed services learned in the late 1940s and early 1950s, determined which services would receive funding for the development and procurement of new weapons. For this reason, the doctrinal debates within the armed services often developed into heated exchanges. Moreover, civilian intervention in targeting debates, for example, proved difficult and time-consuming because the military organizations had the expertise and controlled much of the information. Eisenhower's decision to retain massive retaliation as part of national strategy inadvertently contributed to the problems of formulating nuclear strategy. The policy of massive retaliation provided no guidance in determining the object of nuclear strikes or the adequate force size. Thus, nuclear strategy debates could never be put to rest. Its ambiguity on these points meant that they were continually raised and debated.

President Eisenhower's views on massive retaliation during the missile gap period reveal more consistency and logic than has generally been recognized. Scholars have been unable to reconcile the Eisenhower who was committed to the policy of massive retaliation, issued nuclear threats in crises, and increased nuclear forces with the Eisenhower who deplored the costs of the arms race and modern warfare, and strictly limited military preparations for the employment of nuclear weapons in crises. Many scholars, particularly Eisenhower revisionists, select one as the "real" Eisenhower, dismissing the other as having been conjured up for political consumption or for some other reason. But in fact, Eisenhower espoused and integrated these two seemingly different sets of beliefs.

Scholars have also tended to underestimate the significance of debates over national strategy and, to a lesser extent, nuclear strategy during the last three years of Eisenhower's second term. However, declassified documents reveal a lengthy debate within the Eisenhower administration over both national strategy and nuclear strategy. Eisenhower and his aides confronted the problems of nuclear deterrence in an age of mutual vulnerability. They began the movement away from massive retaliation and established the foundation for MAD. President Eisenhower's goals and the authority he wielded over policy making were the major reasons why more change did not occur during these years.

Massive Retaliation as a National Strategy

At the time of Sputnik, massive retaliation was the central element of the Eisenhower administration's national strategy. Despite public criticism and occasional statements by the administration, U.S. policy called for the employment of nuclear weapons if either extended deterrence or strategic deterrence should fail. The BNSP paper in effect in October 1957, NSC 5707/8, stated: "It is the policy of the United States to place main, but not sole, reliance on nuclear weapons; to integrate nuclear weapons with other weapons in the arsenal of the United States; to consider them as conventional weapons from a military point of view; and to use them when required to achieve national objectives. Advance authorization for their use is as determined by the President."[1] This paragraph retained the spirit—if not the language—of the New Look as first enunciated in 1953, in NSC 162/2.[2] But changes in the external environment brought new concerns about the wisdom of massive retaliation. Could the United States rely on using nuclear weapons to support its foreign policy objectives once Soviet missiles could retaliate against the United States? Would Soviet missiles negate U.S. nuclear power, allowing the U.S.S.R. to conduct limited aggression in Berlin or the Third World? Should the United States make nuclear threats in subsequent crises as it had in the past? What effect would the continuation of this policy have on the Allies? The British, who were on the verge of joining the thermonuclear club, had already cited U.S. vulnerability and the declining credibility of massive retaliation as reasons for their pursuit of an independent deterrent.[3]

The immediate post-Sputnik uproar over U.S. military policy was, for the most part, precipitated by weapons programs rather than by national strategy. The Gaither report, with its alarmist calls for strengthening deterrence with a $44 billion increase in defense spending, gave President Eisenhower no advice on which national and nuclear strategies would best deter the Soviet Union. This was not a mere oversight. The panel did not include such recommendations because it could not agree on four points, including the credibility of nuclear deterrence and reliance on nuclear weapons in the U.S. defense posture. Paul Nitze, who had been brought in to help write the Gaither report, explained each point in depth.[4] Nitze believed that the threat of nuclear retaliation would be a credible defense against fewer threats once the Soviets had deployed missiles. He cited two countervailing pressures that affected American employment of threats to achieve political goals: "As we see Russian nuclear offensive and defensive capabilities increase we tend to decrease the range of possibilities it seems prudent to cover through nuclear deterrence. As we study the high cost and difficulty of non-nuclear means of handling the various possible Soviet-Communist middle-range threats we tend to expand the range of issues we feel justified in covering through nuclear deterrence." Nitze concluded that "one is led to the view that a severe restriction of the range of issues to be covered by general nuclear deterrence will be called for in the period beginning in 1959." Some panel members suggested that the aim of U.S. nuclear policy should be to deter only the first use of nuclear weapons by other nations. Nonnuclear communist attacks could be deterred through collective efforts and "the unilateral right to use nuclear weapons within one's own country in self-defense." Another question dividing the panel, according to Nitze, was the U.S. "emphasis" on nuclear weapons in foreign policy. In a repudiation of massive retaliation and the New Look, Nitze argued that the United States might achieve more in international affairs by employing other instruments (political, economic, and conventional military forces) than by relying on nuclear weapons. Nuclear deterrence would continue to be an important element in U.S. defense policy, but an overreliance on it and "public concentration on the possibility of nuclear war hurts us in the cold war rather than helps us."[5] For some Gaither panel members and for Nitze, the emphasis on the nuclear shield in U.S. foreign policy was coming to an end because of the threat of Soviet missiles.

Nitze's memorandum raised the nuclear policy questions that the Gaither panel failed to resolve. The Gaither report endorsed an across-the-board expansion of nuclear forces but remained silent on the objectives which should guide nuclear policy and strategy. It addressed massive retaliation only indirectly in the laundry list of weapons programs it offered the president. The panel's failure to examine these important issues constitutes a serious deficiency. Several items, such as increased spending for limited war forces, were inconsistent with massive retaliation. The problems of nuclear deterrence that the panel did not address were the most contentious and bothersome defense issues in the remainder of the Eisenhower presidency.

The first post-Sputnik discussions of massive retaliation as national strategy took place during the administration's annual review of BNSP. The BNSP paper matched the means with the threats to guide the pursuit of national objectives. The long paper held great importance because it established planning priorities—and thus budget priorities—for the coming year. Development of the BNSP was one of the most important functions of the NSC during the Eisenhower years, and the task fell to its Planning Board, the interagency board of deputies then chaired by Eisenhower's special assistant for national security affairs, Robert Cutler. Even before the Sputnik launch, Cutler worried that the BNSP policy on nuclear weapons lacked credibility, especially in the event of limited conventional conflicts.[6] A JCS war game of an ICBM exchange in the 1960s completed in late 1957 reinforced Cutler's views. It found that neither superpower could limit "the extreme magnitude and rate of losses quickly sustained." In the future, the United States might have to limit damage by employing "clean" nuclear warheads that released lower amounts of radiation thus allowing more "selective" targeting.[7] The JCS war game reinforced Cutler's belief that U.S. strategy needed revision because the unwillingness of the United States to use nuclear weapons would embolden the Soviets and decrease cohesion among the NATO members. After receiving a JCS briefing in January 1958, Cutler arranged a similar one for the president and other senior foreign policy makers.

Cutler seized upon the potential of clean weapons to cause fewer civilian casualties as a means to restore credibility in nuclear threats when the AEC commissioners pressed Eisenhower to approve clean weapons development in early 1958.[8] Navy Captain Jack Morse, the AEC's representative to the NSC, provided Cutler with arguments in support of clean weapons, just as he had in August 1957.[9] Morse asserted that because of the threat of casualties to the U.S. population, "neither Russia nor our allies believe that we will invoke all-out war short of direct attack, and it is past time for us to recognize that fact ourselves."[10] By the spring of 1958, Cutler had begun to develop a coalition in support of challenging the policy of massive retaliation.

On March 20, 1958, Cutler raised the problems of massive retaliation with the president and other senior officials in meetings both with the NSC and in the Oval Office. He had the head of the Federal Civil Defense Administration brief the NSC on the potential effects of dirty and clean nuclear fallout in the United States.[11] Cutler also sought NSC guidance with regard to BNSP development and outlined what the role of intelligence estimates should be in the process. A still secret NIE section entitled "The State of Mutual Deterrence and the Deterioration of the Western Position," "disturbed" Cutler. He asked John Foster Dulles, "What are we going to do about the fear of our allies that the United States will not use its nuclear retaliatory capability to protect allies from Soviet aggression?" Dulles said that he "could not understand what so concerned General Cutler." He did not think "that our allies were losing faith in our will to make use of our nuclear retaliatory capability in the event of a Soviet attack." But Cutler was not convinced. Dulles explained that: "mutual deterrence would not only apply to large wars but, to some degree at least, it would also apply to little wars. Did General Cutler object to this situation? What was wrong with mutual deterrence? Did General Cutler advocate war?" Cutler answered that the Soviets would use mutual deterrence to undermine the West by means of limited aggression using conventional forces. The president and Secretary of State Dulles rejected the idea that the Soviets would take such actions, but Cutler still would not budge, arguing that "we [do] not have conventional forces available" to defeat Soviet conventional forces.[12]

Later that afternoon, Cutler, Goodpaster, and several other top officials met with the president to discuss clean weapons and U.S. targeting policy (see the next section). Cutler asked whether nuclear testing should be continued in order to develop a class of large, clean nuclear weapons. He made the case for clean weapons on the basis of the declining credibility of extended deterrence:

> If the U.S. loses its will to retaliate when certain areas other than its own territory are violated, its alliances will crumble and the enemy will be emboldened to take greater risks in subversion, economic penetration, even in minor military aggression on outlying or minor areas. . . . When a state of mutual deterrence has arrived, the U.S. will eventually be forced, openly or tacitly, to admit that we will never use our massive retaliatory power with "dirty" weapons *except* in self-defense. By that admission, the U.S. will have lost the *element of uncertainty* as to retaliation against attack other than on the homeland, which is the core of "deterrent value."

Large warheads 95 percent clean could enable the United States to use nuclear weapons but inflict fewer civilian casualties, perhaps avoid automatic escalation to general war, and restore some of massive retaliation's lost credibility. The members of the group decided to continue the development of clean weapons, although they could not agree on the "implications" of these clean weapons. After the meeting, Cutler recorded his thoughts as to why his case was not received more favorably, among the reasons: "*We* would not use clean weapons in connection with massive retaliation—and neither would Russians, who care nothing for lives."[13]

Those making the case for clean nuclear weapons suffered additional setbacks. Scientific advisers questioned the degree of "cleanliness" that could be achieved in nuclear weapons.[14] Cutler was rebuffed when he reintroduced the subject of clean weapons in subsequent NSC discussions. Hoegh, head of the Federal Civil Defense Administration, told the council that simulated attacks indicated that there would be about 21 million fewer United States casualties if the Soviets employed clean weapons than if they used dirty weapons (30.4 million and 51.3 million casualties, respectively), but President Eisenhower said he doubted the utility of such estimates: "If . . . we are thinking of an attack which would involve some 30 million U.S. casualties in the initial exchange, we have still not approached the casualty limits." He believed that the final number of casualties would be "impossible" to estimate.[15]

With a BNSP draft only weeks away, Cutler had established credibility as the major issue with regard to national strategy. In NSC meetings, Planning Board discussions, and private correspondence, he pressed the Eisenhower administration to reexamine massive retaliation on the grounds of declining credibility. Cutler succeeded in convincing senior administration officials that a change was needed, even though his solution—clean nuclear weapons—failed to gain much support. He had put the issue on the agenda and had begun to build a coalition, but he needed the support of the most influential administration officials, primarily those in the Defense and State Departments, if he was to have a chance of convincing Eisenhower.

Soon after the March 1958 meetings, John Foster Dulles began to question the utility of massive retaliation. It is unclear whether he had any doubts about it prior to the March meetings.[16] But in their meeting of April 1, Dulles and Eisenhower made clear that they shared Cutler's concerns. The secretary said that U.S. policy "too much invoked massive nuclear attack in the event of any clash anywhere of U.S. [*sic*] with Soviet forces." He wanted a review of massive retaliation and the possible alternatives—such as tactical nuclear weapons—that could be employed without threatening the Soviets with "wholesale obliteration." Dulles explained that massive retaliation limited U.S. policy: "I pointed out that there was a certain vicious circle in that so long as the strategic concept contemplated this, our arsenal of weapons had to be adapted primarily to that purpose and so long as our arsenal of weapons was adequate only for that kind of a response, we were compelled to rely on that kind of response." Any changes would have to be made "carefully," Dulles confided, since an admission of the availability of options short of massive retaliation would "weaken" deterrence. Eisenhower agreed, saying that "he, too, was under the impression that our strategic concept did not adequately take account of the possibilities of limited war." The two decided that the problem should be reviewed by senior policy makers.[17]

Dulles and Defense Secretary McElroy discussed the future of massive retaliation with the service secretaries, the JCS, AEC Chairman Lewis Strauss, and Cutler on April 7. Secretary Dulles said that "new conditions are emerging which do not invalidate the massive retaliation concept, but put limitations on it and require it to be supplemented by other measures." Continuing the strategy without adjustment would also put considerable strain the NATO alliance. "Our allies are beginning to show doubt as to whether we would in fact use our H-weapons if we were not ourselves attacked. . . . It is State's considered opinion that although we can hold our alliance together for another year or so, we cannot expect to do so beyond that time on the basis of our present concept. Accordingly, we should be trying to find an alternative possessing greater credibility." Defense Department representatives sympathized with Dulles but had different perspectives, reflecting their organizational affiliation. The Air Force officers (CJCS Twining and CSUSAF White) complained that the United State already put too much emphasis on tactical nuclear weapons; Chief of Naval Operations (CNO) Arleigh Burke and CSUSA Maxwell Taylor argued for an increase in nonstrategic forces, both conventional and nuclear. Defense Secretary McElroy wondered if tactical nuclear weapons could be employed without escalating to general war.[18] The meeting concluded without a consensus, in part because the State and Defense Departments had not yet completed a study of limited war forces that had been prompted by the Gaither report.

After the meeting, Cutler listed ten factors influencing national strategy and sent the memorandum to Dulles and McElroy. Entitled "Some Elements for a Realistic Military Strategy in a Time of Maximum Tension and Distrust," it indicates the recognition that massive retaliation was of limited value and reflects many of the concepts contained in the policy of mutual assured destruction. It is quoted here in its entirety except for a brief passage that has been sanitized.

1. General war is obsolete, because of its incalculable destructiveness, as a method to obtain national objectives.

2. The U.S. will not launch "preventive war."

3. The original purpose of our maintaining a massive nuclear capability to wage general war, through immediate retaliation, is to deter a hostile power from aggression against the U.S., U.S. forces, and the allies to whose defense the U.S. is committed. (Doubt is growing in many areas whether U.S. nuclear retaliatory power would be used *except* against attack on the U.S. and U.S. forces.)

4. Because U.S. nuclear capability is intended for retaliation, not initial attack, the U.S. targeting plan should be based on paralyzing the Soviet nation [SANITIZED] and not on knocking out her war-making capabilities (already launched in large part) at several thousand military targets.

5. When both sides have the nuclear capability substantially to destroy each other, whichever strikes first, the primary use of U.S. resources for defense should be to have ready and invulnerable that capability in sufficient strength; but the U.S. should not devote resources to building up superfluous deterrent capability at the expense of other necessary capabilities and national needs.

6. Strategic nuclear capability is not effectively usable against, or in reply to, minor aggression.

7. By eliminating from the strategic deterrent capability vulnerable elements and elements intended to blunt a once-launched enemy attack, rather than intended to deter its launching, resources could become available to the U.S. effectively to deal in future times with minor aggression and with Communist economic and political penetration overseas.

8. In dealing with limited aggression, the U.S. objective should be to *stabilize* the situation rather than, by pressing for outright victory, to provoke a hostile response which may through counteractions lead on to general war.

9. The building up of U.S. strategic deterrent forces and overseas bases beyond the objective stated in 4 above is as dangerous a provocation to hostile action as not maintaining enough.

10. A large-scale program at high priority by either side can appear a ground for retaliation. Gradual, long-term programs are preferable in this period of tension.[19]

Cutler, echoing the policy's critics, had emphasized the problems of massive retaliation with respect to credibility, force size, and extended deterrence.[20] A national strategy based on massive retaliation seemed a dangerous choice for the missile gap period. But Cutler still had not developed an alternative strategic doctrine that would have the support of a coalition.

On April 15, 1958, the Planning Board circulated a draft paper on BNSP, after fourteen meetings over a two-month period.[21] Designated

NSC 5810, the paper labelled the Sino-Soviet quest for world domination as the "basic threat" to U.S. security. It found that four factors exacerbated the threat: (1) the increasing destructiveness of nuclear weapons; (2) growing Western doubts about the U.S. resolve to use nuclear weapons; (3) unrest in the Third World; and (4) the American public's failure "to appreciate the extent of the crisis facing the United States." The "basic problem" faced by the United States was to deter both general and limited war without ruining the economy while continuing to lead the West. In NSC 5810 the Planning Board continued the New Look reliance on nuclear weapons, repeating verbatim the policy as stated in NSC 5707/8.[22] The board could not agree on a new policy statement on the subject of limited war so it simply retained the language of NSC 5707/8, which called for the use of conventional and nuclear forces to defeat limited aggression in the Third World without escalation to a general war. Cutler, determined to have an NSC discussion of the credibility issue, decided to use the section of the paper on limited war as the basis for his assault on massive retaliation.[23] Contrary to a majority of the Planning Board, he drafted an alternative version of one paragraph, pointing out the importance of limited-war forces in "a period of relative nuclear parity." It stated in part:

> In the case of any such limited military aggression, the United States should decide whether: (1) vital U.S. interests require the defeat of such limited aggression by prompt and resolute application of whatever force is necessary, even at the risk that major Communist counter-action may spread the hostilities into general war, or whether (2) U.S. interests would be served by the application of only the degree of force necessary to achieve objectives of limiting the area and scope of the hostilities and restoring the *status quo ante,* thereby seeking to minimize the risk that major Communist counter-action would spread the hostilities into general war.[24]

Cutler expressed doubt about the viability of massive retaliation and evoked images of future stalemates like the Korean War.

But NSC 5810 provoked strong reactions from the Defense Department. W. J. McNeil, the DOD comptroller, said that it maintained the "containment psychology" but included no long-term, positive program for achieving national security.[25] The JCS divided over Cutler's limited war proposal. The Army, Navy, and Marine Corps endorsed Cutler's new policy language; the Air Force suggested keeping the language of NSC 5707/8 until the State and Defense Departments completed a joint study on the subject of limited war. In a memo prepared for McElroy, CJCS Twining rejected Cutler's language: "This wording would water down our resolution to strike back against Russia in any aggression, and would indicate that we are willing to conduct limited wars without using the force required. This intent would probably become known to the U.S.S.R. and would be an invitation to start limited wars."[26] Twining recommended that the old policy be continued without change and that there be no delay to await the results of the joint study.

The NSC discussed NSC 5810 at its 364th meeting on May 1, 1958. Cutler opened the discussion by outlining the major influences on the review of national security policy, which included:

> *First.* The realization that both sides are capable of delivering massive nuclear devastation (regardless of which side strikes first) increasingly deters each side from initiating, or taking actions which directly risk, general nuclear war.
> *Second.* During this time of nuclear parity and mutual deterrence: (a) there is growing doubt in the Free World whether the United States will use its massive nuclear capability, except in retaliation to direct attack on the United States or its forces, leading to a growth of neutralism and a weakening of Free World alliances; (b) the U.S.S.R. will be more bold. . . .
> *Fifth.* A U.S. massive nuclear retaliatory capability, invulnerable and sufficient to deter general nuclear war, and to prevail in such a war if it comes.
> *Sixth.* A U.S. flexible and selective capability (including nuclear) to deter or suppress limited military aggression; realizing that the chances of keeping a conflict limited—whenever major areas or causes are involved—are at best not promising.[27]

He then introduced his alternative paragraph on limited war, claiming that it would "ensure that the United States would have a flexible capability." Secretary McElroy and his deputy, Donald Quarles, immediately attacked Cutler's limited-war proposal, using rhetoric certain to resonate with the president. McElroy warned that it had "grave potential budget implications." The only way the United States could afford to expand conventional forces would be through reductions in nuclear forces. Quarles argued that since the United States could not preclude the use of nuclear weapons, it "must, on the contrary, rely upon them." Abandoning massive retaliation posed other risks, according to

Quarles: "The danger of speaking about a limited war involving the United States and the U.S.S.R. is precisely that it would encourage this kind of erroneous thinking. It would be extremely dangerous, for example, to allow a concept to get out that if we were attacked in Berlin we would not apply all the necessary military force required to repel the attack. Any other concept than this . . . would have the effect of inviting an attack."[28] Army Chief of Staff Taylor, speaking for the Navy and Marine Corps as well, recommended "immediate" approval of the alternative paragraph and the expansion of limited-war forces. This would offer the United States choices beyond "a massive nuclear strike or simply by retreating in the face of the aggression." Taylor also claimed that mutual deterrence made limited war possible in Europe—not just in the Third World. Consequently, the United States should develop the capability to defeat Soviet conventional forces without relying on nuclear weapons. Both Air Force Chief of Staff White and CJCS Twining opposed Cutler's alternative paragraph on the grounds that U.S. forces for use in a limited war were "strong" and "reasonably adequate," as well as for "psychological" reasons. Emphasizing limited-war forces, Twining advised, would have "an extremely adverse effect" on NATO and would undermine the ability of the United States to deter the Soviets. He termed a reduction of strategic nuclear forces "unacceptable" and "fatal."

When Twining finished, John Foster Dulles, the man most closely associated with massive retaliation, spoke at length. In two or three years, European nations either would "demand" support from the United States in the form of conventional forces or would withdraw from NATO once they had determined that the United States would not employ nuclear weapons in a European conflict.[29] In a remarkably abrupt reversal, Dulles advocated developing clean nuclear weapons "so that we can devise a new strategic concept which will serve to maintain our allies." But this transition could not take place until the United States prepared a new strategy. "The massive nuclear deterrent was running its course as the principal element in our military arsenal, and very great emphasis must be placed on the elements which in the next two or three years can replace massive nuclear retaliatory capability. In short, the United States must be in a position to fight defensive wars which do not involve the total defeat of the enemy." Dulles said he believed that this change should be implemented even if it meant violating the administration's strict fiscal policies.

President Eisenhower, who had been quiet during the meeting, attacked Cutler's assumption that mutual deterrence provided an "umbrella" that made it possible to fight limited wars without the danger of nuclear escalation. "Actually, the umbrella would be a lightning rod. Each small war makes global war more likely," Eisenhower told the NSC. Expanding conventional forces required either a reduction in nuclear weapons or a budget increase that would lead to the establishment of "a garrison state." He also said he believed that a limited war in Europe would escalate to nuclear war and that "he would not want to be the one to withhold resort to the use of nuclear weapons if the Soviets attacked" NATO members. John Foster Dulles agreed but warned that without a defensive capacity "by other means than our resort to massive nuclear retaliation, we would lose our allies."

Eisenhower asked "What else [have we] been trying to do these last years but try to induce our allies to provide themselves with just such a local defensive capability and, moreover, doing our best to help them achieve such a capability?"

The secretary mentioned modernizing military forces to at least give NATO "the illusion" of defense without risking general war. A "bewildered" president responded: "What possibility was there . . . that facing 175 Soviet divisions, well armed both with conventional and nuclear weapons, that our six divisions together with the NATO divisions could oppose such a vast force in a limited war in Europe with the Soviets?" The disagreement between them resurfaced later in the meeting. Referring to his impending trip to Berlin, Dulles said that

> when he [Dulles] got there he would repeat what he had said in Berlin four years ago—namely, that an attack on Berlin would be considered by us to be an attack on the United States. Secretary Dulles added that he did not know whether he himself quite believed this or, indeed, whether his audience would believe it. But he was going to perform this ritual act. The President expressed surprise, and said that if we did not respond in this fashion to a Soviet attack on Berlin, we would first lose the city itself and, shortly after, all of Western Europe.

At the end of the meeting, Eisenhower approved NSC 5810 except for the paragraphs on limited war. The old language in these sections would be retained pending receipt of the joint study by the State and Defense Departments.

The NSC meeting revealed the major divisions within the administration. Eisenhower opposed changing the policy on limited war because to do so might weaken strategic deterrence and increase defense expenditures.[30] The president stated his position forcefully but, in leaving the policy open, he acknowledged the differences of opinion. Civilian Defense Department officials, CJCS Twining, and Air Force officials made clear that they endorsed Eisenhower's views, stressing the budgetary and strategic effects of abandoning massive retaliation. Besides Cutler, only the Army and Navy supported a new national strategy, but their positions were given little credence because their organizational interests had been stated numerous times since the New Look was approved in 1953. John Foster Dulles stood between the two groups; he recognized that problems would be created in Europe by a continued emphasis on massive retaliation but wanted a new strategy clarified before replacing the old one.

Senior administration officials spent May and June 1958 attempting to reach a consensus on massive retaliation and limited war. The Air Force, which had the most to lose from a change in policy, prepared a SAC briefing in which it "justified" the need for large forces and massive retaliation. Air Force Chief of Staff Thomas White remarked: "This is something that I am sure we can win if we put on a good story."[31] During his trip to Europe, John Foster Dulles cabled the president that "our European friends will not in fact depend upon our willingness to initiate general nuclear war if there is an attack on Europe." The United States needed to expand tactical nuclear capabilities instead of strategic weapons. Otherwise: "unless there seems to be some alternative they will turn to pacifism and neutralism. The alternative may not in fact prevent the war from becoming general nuclear war, but may prevent neutralism."[32]

The most important discussions followed the release of a draft of the joint study by the Defense and State Departments in mid-June. In it they recommended increasing some logistic support for use in limited war, better notification of the Allies, and a campaign to inform the American public about the role of nuclear weapons in a limited war. These recommendations, if implemented, would not require much of a change in national policy. On June 17, McElroy and Quarles met with the JCS, the secretaries of the armed services, and John Foster Dulles to discuss the joint study. The U.S. limited-war policy, Dulles told the group, remained adequate in eleven of the twelve potential trouble spots (Iran was the exception). The Soviets probably would not initiate limited war in Europe, but Dulles feared an increasing trend toward neutralism and independence in those nations: "Time is soon coming when our NATO allies will not be satisfied that American will surely go to general nuclear war to defend them, if attacked, and risk American devastation; and will demand a surer strategic concept." If the United States did not formulate a new strategy, Dulles said, these psychological pressures might spur Western European nations to develop a larger deterrent, either independently or in concert. This would produce a "dangerous situation if we do not have a common strategic concept to use in defending our common alliance." Dulles could not offer a solution to the problem, however. One possibility, which received Army Chief of Staff Taylor's endorsement, was an increased role for European nations in policy and control over U.S. nuclear weapons in NATO.[33]

The next day, McElroy recommended to the NSC that nuclear weapons remain the "highest priority," concluding that limited war forces required no immediate changes. The administration should not alter massive retaliation, since "It is our [the Department of Defense's] considered opinion that war with the U.S.S.R. cannot be held to limited operations and limited objectives. Moreover, to imply that we might seek to hold a war with the U.S.S.R. to limited operations and limited objectives would involve a dangerous weakening of our deterrent position and certainly have a deleterious effect on the attitude of our allies."[34] The NSC reviewed the joint study at its meeting on June 26 but there was almost no substantive discussion.[35] The civilian officials in the Defense Department had successfully defended massive retaliation against Cutler's attack by claiming that conventional forces were adequate and that an alternative did not exist. Cutler's bureaucratic politicking succeeded until the last stages of the debate, but none of the four dominant actors—Eisenhower, Dulles, McElroy, and Twining—offered unconditional support. John Foster Dulles did not like the conclusions of the joint study, but he decided not to challenge it because he still had not developed an alternative to the policy of massive retaliation.

All that remained was final ratification of NSC 5810 by the president and the NSC. On the day before the NSC met, John Foster Dulles confided to Eisenhower that massive retaliation "is rapidly outliving its usefulness and we need to apply ourselves urgently to finding an alternative strategic concept. I [do] not, however, wish to air my misgivings on this sensitive subject before the Council."[36] At the NSC meeting, on July 24, Dulles said he supported continuing the old language of NSC 5707/8 for budgetary reasons, adding that the policy of massive retaliation "will not be fixed for all time." Thus, the problem of massive retaliation's lack of credibility in Europe pushed John Foster Dulles to the brink of endorsing a policy of Flexible Response, only to have him back down because of economic costs, Eisenhower's obstinance, and his own inability to present an acceptable alternative. But his position seems to have had an effect on the president. Although Eisenhower approved NSC 5810/1 as the official statement of BNSP, he wanted the policy on limited war "kept open" to allow for additional examination. The Defense Department, Eisenhower stated, should develop its budget "on the basis of the old language."[37]

The formulation of BNSP in 1958 was the most in-depth examination of national strategy that was made during the missile gap period. Cutler, influenced by the AEC and the Army and Navy, pushed for a major revision but failed to achieve it because of opposition and doubt on the part of the president. Eisenhower retained massive retaliation for purposes of budgetary control, but the debate had influenced him. Contrary to the conventional wisdom, Eisenhower remained more strongly committed to massive retaliation than John Foster Dulles, in part because each weighed alliance members' behavior, strategic deterrence, and fiscal effects differently.

Eisenhower's decision to leave the policy on limited war under review meant that the issue simmered throughout the remainder of 1958 and into 1959. As the annual BNSP review for 1959 approached, the political constellations in the administration had realigned significantly. Both Robert Cutler and John Foster Dulles had left the administration. Even though their successors—Gordon Gray and Christian Herter, respectively—shared their concerns about massive retaliation, Dulles's stature and Cutler's skills in managing the policy process were irreplaceable. Thus, although the debate over the policy of limited war was replayed in 1959, it proved to be a weak imitation of the 1958 debate, that had little chance of overturning massive retaliation.

In preparation for the review, the State Department pressed for a joint study on limited war, to be conducted in conjunction with the DOD, in the hope that its results would influence the language in the BNSP paper. But the State Department's efforts collapsed when McElroy discovered its scope and potential effect. The Defense Department continued to oppose any changes in BNSP that would result in an increase in limited war forces at the expense of strategic forces. Former CJCS Arthur Radford criticized the State Department for believing that a large war could be fought without the danger of nuclear escalation. "Assistant Secretary of State Gerard Smith, [Radford] said, is attempting to get a change of policy, even though Herter says that he is seeking nothing so sweeping."[38]

The NSC Planning Board drafted a new paper on BNSP (NSC 5906) based on two NIEs, advice from twenty-two consultants, and the results of eleven meetings. The draft contained few changes relating to massive retaliation, even though much of the paper had been revised. One interesting modification determined that a change in the military balance would constitute a "basic threat to the U.S."[39] As in the previous year, Gray prohibited the Planning Board from discussing limited war in the draft because of the sensitive nature of the controversy. He tried to resolve the dispute by conducting direct negotiations between the president, McElroy, and Herter. He agreed with the State Department's position: "I told the President that I felt it was not wise to ignore the repeated pressures, both within government and out, for an examination of the limited war problem." But the president again emphasized budgetary considerations in national policy formulation in its early stages; he questioned whether "these people who were raising the clamor really were ready to establish priorities to do these things without resorting to regimentation."[40] Frustrated by the slow pace of negotiations, the State Department resurrected the alternative paragraph on limited war that Cutler had proposed in the previous year and circulated it as an annex to NSC 5906.[41]

The draft parper on BNSP dominated the summer agendas of national security policy makers. The NSC devoted part of five consecutive meetings to it, and it was the subject of numerous Planning Board meetings and conferences between President Eisenhower and Gordon Gray. The discussions concentrated on the sections on limited war forces and a second, still classified section (paragraph 12) on policy concerning the use of nuclear weapons. On July 15, Gray showed his proposed amendments to paragraph 12 to the president. He recorded:

> I said that it seemed to me now clear at the Planning Board level the State Department view was that we do not *now* have an adequate conventional limited war capability and that the State Department effort would be to enlarge our *present* capabilities. . . . I then pointed out to him that the

> Defense Department felt that the only criteria that should govern the use of nuclear weapons were military criteria and that the Defense Department would want 12*a* to read: "Planning should contemplate situations short of general war where the use of nuclear weapons manifestly not be militarily necessary nor militarily appropriate to the accomplishment of national objectives."[42]

Gray also worked closely with the president to revise a paragraph on limited war so that it would apply only in non-NATO conflicts and would "set the stage for planning" in the Defense and State Departments. Eisenhower allowed such marginal changes to be made in limited war policy because "a change in language is not necessarily a change in policy."[43] Gray, unlike Cutler, appeared ready to negotiate small adjustments in language that seemed to moderate the policy of massive retaliation but represented no real change in national policy.

The NSC discussed the BNSP paper's sections on limited war on July 9 and 30. On both occasions, Secretary of State Herter sought further clarity concerning the use of nuclear weapons in a limited war, only to be rebuffed by the president and officials of the Defense Department. Herter succeeded only in gaining Eisenhower's assurances that the policy would not lead to an overreliance on nuclear weapons. The president rejected the State Department's proposal for developing "balanced" military forces for use against the Soviets.[44] In fact, Eisenhower made his policy preferences clear at the beginning of the debate, when he told the NSC that he did not expect any "radical change in policy."[45] On August 5, 1959, President Eisenhower officially approved NSC 5906/1—the last BNSP paper issued during his presidency. To protect his priorities, the president directed that "final determination on budget requests based [on NSC 5906/1] will be made by the President after normal budgetary review." Thus the BNSP paper could not be used to automatically drive military spending higher by establishing new policy priorities or requirements.[46]

The formulation of BNSP in 1959 taxed the president's patience, even though the debates were less contentious than those of the previous year. He was particularly irritated by the inability of his senior aides to reach a consensus on issues concerning the policy on limited war. Gray recorded that Eisenhower "expressed his displeasure at not being able to find language which was clear and decisive and would communicate to everyone his clear intention." Clarity of language in NSC 5906/1 concerned Eisenhower and Gray more than usual because neither expected to revise BNSP in 1960.[47] The president expressed his dissatisfaction to the NSC on more than one occasion during these months.[48]

In the debates over national strategy in 1958 and 1959, the Eisenhower administration acknowledged the problems raised by limited war policy but refused to implement any substantial changes. Massive retaliation remained the central element of national security policy despite contentions that Soviet strategic forces nullified the credibility of that doctrine. There are at least two reasons for the doctrine's resilience. First, the administration believed that massive retaliation offered the least costly and most effective way to deter general war. A strategy that put greater emphasis on conventional forces would require an immediate and dramatic increase in the defense budget. Second, Eisenhower and DOD officials worried that a de-emphasis on nuclear weapons might encourage Soviet aggression or even result in the failure of strategic deterrence. But each year, there was significant opposition within the administration to continuing a national strategy based on massive retaliation. That massive retaliation survived is testament to President Eisenhower's skill and his determination to control the formulation of national strategy.

President Eisenhower's reasons for continuing massive retaliation ran deeper, as revealed by the 1959 debate over the U.S. objectives in the event of a general war with the Soviet Union. Since 1948, the NSC had periodically revised the paper on U.S. general war objectives, most recently in 1954 with the approval of NSC 5410/1. In January 1958, Cutler initiated a review of that paper in conjunction with the annual BNSP review.[49] This would constitute the most comprehensive review of national strategy since the New Look. But the long BNSP debate in 1958, combined with Cutler's departure from the administration and the time spent on budget formulation, delayed review of the paper on general war objectives until early 1959.

On January 22, 1959, Gordon Gray sought NSC guidance for a new policy by circulating a discussion paper. The issue of general war objectives held special significance for national security policy because of the Soviet Union's May 27 deadline for the withdrawal of Western troops from Berlin. Formulating war objectives antebellum is a complicated undertaking and few members of the NSC welcomed it. Gray said revision of the paper was the "most difficult task" he had faced since replacing Cutler; John Foster Dulles said "he would personally hate" having to do it. Some Planning Board members thought that NSC 5410/1 should be "rescinded" and not replaced at all. But the council agreed with Dulles that they should at least try to write a new policy statement. Gray asked whether it should apply to attacks by communist bloc nations other than the Soviet Union and to limited wars. Further, should the United States include Communist China when planning its retaliation against possible Soviet attacks? Dulles and McElroy asserted that the paper should be limited to general war because formulating a policy that would be applicable to all situations would be too difficult. Dulles wanted further examination of targeting China and offered the observation that "The assumption . . . that you could have a general nuclear war in which a 'victory' could be achieved also needed to be reconsidered."

Gray then introduced the issue of war termination. In the discussion paper, he had asked:

> Should the United States in the event of general war initiated by the U.S.S.R.: Despite the loss of U.S. lives and resources which might be involved, endeavor by all necessary means to reduce the capabilities of the U.S.S.R. to the point where it has lost its will or ability to wage war against the United States and its allies; and yet be prepared to consider an offer by the U.S.S.R. to cut short the nuclear exchange at a point advantageous to the United States, even though the U.S.S.R. might retain some will and ability to continue the struggle?[50]

The president interrupted him almost immediately, and reminded the NSC that "with respect to these questions . . . everyone knew that . . . we took Clausewitz as our guide. Clausewitz, in his doctrine, put all his emphasis on the destruction of the will of the enemy to wage war rather than the enemy's capabilities to do so. . . . perhaps it was rather futile to try to make a real differentiation" between the two. The president then translated the theory into policy terms: "War is after all waged for a purpose. Our purpose is to defend ourselves. To defend ourselves means that we must destroy the present threat to ourselves. Accordingly, once we become involved in a nuclear exchange with the Soviet Union, we could not stop until we had finished off the enemy; that is, forced him to stop fighting. If at any point in the hostilities we agree to make terms with the enemy, we would only make terms which allayed the Communist threat to us."

When Gray asked whether the United States should undertake postwar planning (including surrender terms and territorial boundaries), Eisenhower laughed and Dulles smiled at the thought. The president explained that such planning required imagination, but "imagination will not solve non-imaginable problems." When CJCS Twining interrupted, Eisenhower asked him, "What could be our objectives in . . . a general war beyond the objective of hitting the Russians as hard as we could?"

"We planned," Twining answered, "in the event of . . . a war 'to shoot the works' and not to apply our military power by degree." Critical of negotiating with the Soviets during a war, Twining added: "We finish the attack and then talk to them but we shoot the works."

The president also disliked the idea of wartime negotiations. He chillingly and bluntly told the NSC: "The only form which you could expect to get a peace offering would be from that side in the conflict which was putting up the white flag. The U.S. will never do this so we should go ahead and hit the Russians as hard as possible. We could not do anything else. They . . . will have started the war, we will finish it. That is all the policy [I have]."[51] Not surprisingly, this statement effectively capped the NSC's discussion; Eisenhower and the NSC decided that the Planning Board should draft a new paper. But the president had strongly stated his views in terms consistent with massive retaliation. In effect, he gave the Planning Board little leeway in drafting the new policy.

The Planning Board completed a draft policy paper (NSC 5904) in mid-February, with NSC review scheduled for March 5. Much of the first section, on U.S. objectives, had been copied almost identically from NSC 5410/1. In a general war with the Soviet Union, according to NSC 5904, the United States should seek to "prevail" by "reducing" the Sino-Soviet bloc's will and its capacity for war. An attempt should also be made to undermine Soviet domestic political control. Other Eastern bloc nations should be subject to U.S. attack only if they participated in the Soviet attacks. To achieve these objectives, "the United States should utilize all requisite force against selected targets in waging war against" the Sino-Soviet bloc. The draft paper unrealistically raised expectations about the possibility of terminating a general war and achieving postwar objectives—reflecting Eisenhower's views.

The second section of NSC 5904, which dealt with United States objectives in wars not including the Soviets, again focused attention on the credibility of massive retaliation. The paper suggested that before entering hostilities, the United States should evaluate the probability that such an action would precipitate a general war with the Soviet Union. The JCS disagreed with the majority of the NSC Planning Board over the policy that should guide U.S. actions in such a scenario. The board's majority endorsed "terminating hostilities" before achieving all objectives, in order to avoid a general war, whereas the JCS claimed that "once committed, the clear and immediate danger of general war with the U.S.S.R. must not deter the United States from taking the actions necessary to achieve its objectives."[52] The DOD staked out the middle ground, suggesting: "The U.S. will have to decide in the light of the circumstances then existing whether it is in the U.S. interest to pursue its original objectives." These splits mirrored the 1959 divisions over BNSP.

The NSC reviewed the draft on March 5, 1959, just as congressional pressure for increasing U.S. troop strength in Berlin peaked.[53] Christian Herter, the acting secretary of state, argued against including China and other communist states as targets in war plans if they were not belligerents. The JCS, and the Defense and Treasury Departments opposed Herter. The president jumped into the discussion and rejected Herter's assertion that China would be attacked automatically under this policy. Eisenhower's semantical argument cited the paper's call for use of "military and other measures" to justify his interpretation. Besides, he said, "If the U.S. . . . got into a disastrous nuclear war with the Soviet Union and in the course of the war simply ignored Communist China, we would end up in a 'hell of a fix'." Gray said the NSC already agreed that retaliation against China and Eastern Europe should not be indiscriminate. The president concurred, again reminding the NSC of the need to attack China should a war break out: "He said he simply could not envisage the U.S. becoming involved in an all-out nuclear war with the Soviet Union while at the same time permitting Communist China to stay on the sidelines and develop, after perhaps forty years, into another Soviet Union." Eisenhower agreed that the Soviet bloc countries, particularly those in Eastern Europe, should not be subjected to unnecessary destruction. Herter doubted whether China would join the Soviets in a war. But the president again argued that "we simply could not just ignore a Communist China which remained untouched and intact after a terrible war between the U.S. and the U.S.S.R. To do so would be unrealistic in the extreme." McElroy agreed with Eisenhower that China would "certainly" side with the Soviets. The new policy, the president said, should ensure that neither the Soviets nor the Chinese could injure the United States after a war.[54]

The paper as approved, NSC 5904/1, became official policy in mid-March, but only after a series of bureaucratic maneuvers by the State and Defense Departments had caused the NSC to hold another meeting on the draft paper. The final version called for destroying the will and capability to wage war of both the Soviet Union and China but stated that other communist states should have to endure as little nonmilitary destruction as possible. NSC 5904/1 rejected any wartime negotiations unless no postwar threat remained. The paper included a paragraph mildly calling for the reexamination of objectives if, in a war not involving the Soviets, escalation was "a clear probability."[55] NSC 5904/1 continued to be official policy until the end of the Eisenhower administration.[56]

This debate over U.S. objectives in the event of a nuclear war indicates President Eisenhower's strong commitment to the policy of massive retaliation for strategic reasons. Records of his private statements, which were not diluted or otherwise prepared for public (or specifically for Soviet) consumption, reveal a willingness to employ wholesale nuclear strikes against the Soviet Union and Communist China without reservation or interruption. Eisenhower envisioned no alternative in a nuclear war except the total destruction of the Soviet Union since the conflict would be fought to the finish. This is not to say that he was unmindful of the horror of nuclear war or of the implications of such a decision—the public record and declassified documents are replete with Eisenhower's reminders of the overwhelming destruction that can be caused by nuclear war. Rather, his continued adherence to the policy of massive retaliation reflected a resignation that once the two superpowers had begun shooting at each other, escalation could not be controlled and that neither would capitulate in the middle of the war. At times these beliefs were reflected in seemingly contradictory behavior by the president. In crises such as those in Berlin and the Taiwan Straits, Eisenhower publicly stated his determination to use nuclear weapons yet strictly limited the military preparations for such operations. These actions puzzled Eisenhower's military commanders and have confounded scholars as well.[57] Many scholars seem to emphasize one aspect of Eisenhower's behavior and either praise or criticize, as the case may be. Ronald Pruessen, for example, has attacked Eisenhower and Dulles for blithely ignoring the potential consequences of nuclear threats.[58] But Eisenhower's behavior was more logically consistent than his contemporaries realized or that many scholars have recognized. His belief that small wars would, in the absence of restraints and negotiations, escalate to a general war led him to be cautious in planning military operations and to warn potential adversaries of the dangerous outcomes. To Eisenhower, massive retaliation posed fewer risks and imposed fewer costs than expanding limited war capabilities. An increased concentration on preparation for limited war, whether waged with nuclear forces or with conventional forces, could undermine strategic deterrence by allowing the Soviets to believe that the United States would not escalate to a general nuclear war to protect itself and its allies. Replacing massive retaliation also meant incurring higher defense expenditures than the president wanted. Having weighed these costs, Eisenhower decided he could tolerate some decrease in credibility. His continuation of massive retaliation also rested on the strategy's flexibility—it could be adjusted with the changing requirements of the United States. Massive retaliation became essentially a threat to escalate a conflict between the superpowers to a strategic nuclear war that risked the destruction of both the United States and the Soviet Union. Eisenhower clearly believed that only the threat of escalation to a general war could deter the Soviet Union. He reasoned that massive retaliation which threatened mutual destruction remained the best and only effective strategy.

Massive Retaliation as a Nuclear Strategy

The Eisenhower administration struggled to formulate a nuclear strategy at the same time that it was engaged in the debates over national strategy. It needed to decide on a targeting doctrine and an appropriate force posture. Although the administration had addressed such issues in the past, changes in the force postures of both the United States and the Soviet Union meant that they required a fresh look. National strategy and nuclear strategy were intertwined, but the formulation of nuclear strategy had a different dynamic. It involved fewer actors who required more time to reach a decision, and it was less amenable to presidential control. The military services, which played a minor role in the making of national strategy, influenced decision making with regard to nuclear strategy through their control of vital technical information and their responsibility for implementing the nuclear plans. For these reasons, President Eisenhower grew increasingly disgusted with the policy process.

As discussed in the previous section, the Gaither report included no recommendation about nuclear strategy. Paul Nitze, in his examination of unresolved issues, asked whether the United States should program its strategic forces for a fast reaction to a Soviet nuclear attack, or for a slow reaction. He worried the United States might decrease the reaction time of strategic forces as a hedge against a possible attack by Soviet missiles: "A corollary of quick reaction time, however, is that decisions which can involve the world in an irreversible course to general nuclear war must be widely delegated to subordinate units. This involves very great hazards of an unintended initiation of general war and makes normal political relationships impossible. . . . Negotiation with the enemy, use of U.N. procedures, or any of the other steps which even the most rudimentary ethical and political ideas would call for become meaningless." Nitze supported less vulnerable weapons systems, such as Polaris, which promoted civilian control during crises. Targeting doctrine was the most important issue that the Gaither panel had not resolved. In the longest section of his paper, Nitze contrasted the liabilities of two alternative nuclear strategies—"disarming" and "retaliation" against cities. His discussion foreshadowed the debate over counterforce and mutual assured destruction; it also reveals Nitze's personal views at this critical juncture. For these reasons, it is quoted here at length.

> The capabilities necessary to implement a strategy of disarming the enemy differ markedly from the forces which would be appropriate for a strategy of deterrence through maintenance of a secure capability to retaliate. . . .
>
> There is inevitably a degree of conflict between the two strategies and the preparations they call for. With limited resources difficult decisions must be made one way or the other.
>
> There can be little doubt that if we ever get into a war with the U.S.S.R. we would have a much better chance of defeating her and of minimizing the destruction to the United States, its people and its values if we carry out a disarming strategy.
>
> The question is not one of desirability but of practicality. . . .
>
> The danger in a strategy to disarm coupled with a policy which rules out preventive war is that it tends to produce a highly unstable situation between two antagonists. Each must be prepared to strike on a moment's

> notice on any indication that the enemy is about to move. The prospects that such a situation can long endure without being triggered off, perhaps by accidental causes, is not high. There are also substantial cold war political costs to our side from long continued tensions of the type generated by such a situation.
>
> There are, therefore, convinced advocates of putting heavier reliance upon a strategy of deterrence through maintenance of a secure capacity to retaliate even if the maintenance of such capacity means we must cut down our capabilities to conduct a strategy of disarming the enemy.
>
> There can be no doubt but that any war in which we receive the initial blow would be most destructive and almost impossible to win in any meaningful sense. The argument, however, can be made that the chances of nuclear war occurring are much reduced if we have the capability under all circumstances of delivering a highly damaging retaliatory blow at the enemy. If this is to be our strategy we will have to move the cut-off point of what it is we are proposing to deter by our general war nuclear offensive and defensive capabilities over to the dark side of the spectrum of Soviet-Communist actions. We will therefore have heavy requirements in preparing to meet the various threats at the lighter end of the spectrum through non-nuclear means. The question therefore arises as to how big a retaliatory blow we must be able to deliver under all circumstances, including a surprise attack upon S.A.C. . . .
>
> It can be argued that the Soviet leaders would not be apt to follow any course of action which they thought would lead to the destruction of 20 percent or more of Russian population and centers of industry. The chances of the present regime surviving any such disaster would certainly be problematical and would almost certainly be so judged by the men now constituting the Russian leadership.
>
> What would be required to produce such a level of destruction? It can be argued that several hundred megatons delivered on target (if the targets were centers of population and industry) would produce such results. It can also be argued that the ability to deliver two or three thousand megatons more or less at random in the populated areas of Russia would do through fall-out alone.[59]

Nitze believed that a countervalue/minimum deterrence strategy supported by an expansion in conventional forces should replace extended nuclear deterrence. A disarming counterforce strategy posed too many problems and too many risks, in his view, to be a "practical" policy choice. But both strategies required higher defense expenditures and entailed new risks.

Disagreements over nuclear strategy, and targeting doctrine in particular, had been the source of much interservice conflict during the postwar era. The New Look reinforced Air Force control of employment planning and provided the service with budgetary riches. The Army and Navy had tried unsuccessfully to reorient nuclear strategy on a number of occasions. The Navy's targeting doctrine focused on "minimum deterrence," which threatened the Soviets with automatic retaliation against their cities. Sputnik's launch gave the Navy new hope for changing both national strategy and nuclear strategy. In late 1957, the Army and Navy challenged Air Force control of the target lists for nuclear war plans. Through Project Budapest and the "alternative undertakings," they forced the Air Force to develop target lists for use in the event of a debilitating Soviet first strike.[60] The December 1957 JCS ICBM war game provided another boost to the Navy because it demonstrated that the United States could successfully deter the Soviet Union and China "by the threat of the assured loss of a relatively limited number of major vital targets."[61] Navy Captain Jack Morse reinforced Cutler's belief that massive retaliation was no longer a credible strategy when he brought the interservice disputes over targeting to his attention. At Morse's behest, Cutler pressed the president to alter the guidance for the section on U.S. retaliation of the 1958 report by the Net Evaluation Subcommittee, due in November. Morse suggested two alternatives: employing clean weapons only or utilizing a target system limited to the civilian population.[62]

On March 20, 1958, Cutler told the president that current plans to destroy Soviet military targets numbering in the thousands necessitated the explosion of "several million kilotons."

> A recent exercise indicated that, in fifteen hours of preliminary exchange between the aggressor and the U.S., nuclear weapons involving *7 million kilotons* (over half of it in the first three hours) would be detonated; with the U.S. going on to win with still *further* detonations against the enemy. Thus, this exercise contemplated nuclear explosions in North America, Europe, Asia, and North Africa, occurring within a half-day, which were *350,000* times as great in magnitude as the nuclear explosion at Hiroshima.

A nuclear war of these proportions would have an "incalculable" effect on lives, and would perhaps even result in the elimination of life in all forms. This target system might be "reasonable and appropriate in the case of an attack launching *preventive* war," but it would be inappropriate for retaliation because the United States would not be able to destroy thousands of Soviet military targets or to distinguish between expended and unexpended military targets. Consequently, "a *retaliatory* action by the U.S., perhaps 000 hostile targets (1/10 the number above-indicated) would quite as adequately support the concept of deterrence. That is, the enemy would be equally deterred from attacking the U.S., if the enemy knew we would, in retaliation, destroy their 000 population centers instead of only some of their 0000 military installations."[63] Eisenhower approved revising the NESC directive to include "a targeting plan which would seek immediately to paralyze the Russian nation, rather than upon a targeting plan limited to targets of a military character."[64] This plan should decrease the number of both targets and detonations, yet still destroy the Soviets. Cutler told Eisenhower that this was a first step in imposing "strict civilian control over the objectives upon which 'military requirements' for nuclear weapons and forces are based," and it would make clear the costs of each targeting doctrine.[65]

The Eisenhower administration then turned its attention to a strategic force posture that would support the national strategy. The JCS, in response to NSC directives, ordered the WSEG to conduct a study on offensive and defensive weapons and to update a 1957 report entitled "The Relative Military Advantages of Missiles and Manned Military Aircraft."[66] The WSEG completed the latter in early August and forwarded WSEG-23/1 to the administration, although it never briefed the NSC on the report. The group argued that the United States must develop a strategic force comprised of weapons, including ballistic missiles, with different characteristics.[67] This is the basis of the intellectual justification for the strategic triad. The WSEG also advocated two different roles for U.S. strategic forces: assured retaliation and damage limitation. Deterring a Soviet first strike required forces that "would survive to retaliate and inflict unacceptable damage on [the Soviet] homeland." Echoing the "alternative undertakings," the group advocated the targeting of enemy cities and "governmental controls" for retaliatory strikes, since "we must assume that our forces might be reduced to a level such that only this kind of target system could be attacked with assurance of success." But the WSEG preferred that the role of the strategic forces center on damage-limiting counterforce strikes. A counterforce strike against Soviet military forces would "[save] countless U.S. lives and material resources." Counterforce would "better serve our national ends" than would population targeting, but the U.S. response would probably be dictated by the success of the Soviet first strike.

With regard to U.S. strategic force posture, WSEG-23/1 stated that bombers would be the main deterrent "for at least the next five years." ICBMs would eventually provide a "significant" retaliatory capability, the extent of which could be determined once their operational characteristics (in particular, their reliability and delivery accuracy) were better understood. Until such time that the Soviets deployed sizable numbers of ICBMs, the United States would retain a sufficient retaliatory capability by virtue of the ability to launch 350–400 bombers after receiving tactical warning of a Soviet attack.[68] Once the Soviets deployed ICBMs, the United States might lack a credible retaliatory threat until 1962, when it would be able to deploy hardened and dispersed Atlas ICBMs. The Soviet ICBM potential posed problems for counterforce targeting, in WSEG's opinion. Should the United States receive strategic warning of a Soviet attack, "we would place principal emphasis on an attempt to knock out elements of the Soviet nuclear striking power." But counterforce strikes against Soviet ICBMs would be very demanding because of the missiles'quick reaction potential.[69] The U.S. first-generation ICBMs could assist a counterforce targeting strategy only marginally because of the missiles' high CEPs and low reliability. Relatively few targets could be assigned to first-generation missiles. Even by early 1963, the WSEG found,

> the nine Atlas and four Titan squadrons now programmed, totaling 130 missiles, could attack . . . 25 [targets]. The programmed force of five Polaris submarines could attack a further 16 soft targets. The 13 ICBM squadrons and 2 IRBM squadrons based on U.S. soil and the submarine-based Polaris missiles could, under very favorable circumstances, take under attack about 50 soft targets. . . . It is apparent that only a fraction of the counterforce target system could be effectively attacked by first-generation missiles currently programmed and that a counterforce blow in strength will require manned aircraft.

WSEG-23/1 stated that the planned strategic forces must have the capability for staging counterforce missions—an oblique reference to the Navy's preference for minimum deterrence.[70]

WSEG-23/1 indicates the emergence of several important concepts of nuclear strategy. The WSEG advocated a balanced triad of weapons with different characteristics. It recognized, as did Nitze in his memo, that the two missions for nuclear forces might not be equally serviced by the same force posture. Finally, in WSEG-23/1 we begin to see the futility of damage-limiting counterforce strikes.

The WSEG's preparation of the other report (WSEG-30, "Evaluation of Offensive and Defensive Weapons Systems") progressed slowly, even though the administration wished to use it as a guideline in drawing up the Defense Department budget for fiscal year 1960. In WSEG-30 the group examined the entire range of nuclear and conventional weapons systems, assessed the Soviet threat, and analyzed nuclear strategy. Presentation of the report, originally planned for late July, was rescheduled several times. Even though CJCS Twining called WSEG-30 "the most comprehensive and sensitive compilation of weapons and intelligence data ever published as a single document," he twice tried to delay—or even prevent—NSC review of the report. Gordon Gray told President Eisenhower that "General Twining feels that too many items involving 'military hardware' were finding their way on to the NSC agenda."[71] Twining's argument for limiting the NSC's agenda convinced neither the president nor Gray, but they agreed to limit the distribution of WSEG-30 to a small group.[72]

On October 13, 1958, WSEG Director Admiral John H. ("Savvy") Sides briefed a special meeting of the NSC on WSEG-30. After discussing the extent of the Soviet threat and American air defenses, Sides turned to the U.S. strategic nuclear posture and nuclear strategy. WSEG-30, he said, determined the adequacy of strategic forces by their ability "to retaliate in the face of a Soviet attack delivered without strategic warning" against a target list comprised of 135 Soviet cities. Repeating the assessment in WSEG-23/1, Sides claimed that the United States would be able to destroy these targets until mid-1961—or until the Soviets had deployed 100 ICBMs. He made note of a critical period from mid-1960 to 1962, when the United States would not yet have completed all BMEWS or deployed many missiles. Gradually, U.S. missiles would ensure retaliation against the 135 cities, especially as SAC dispersal and missile hardening improved. Sides also mentioned the importance of special qualities of Polaris and Minuteman, which would force the Soviets to expand defensive preparations. In the remainder of the briefing, he touched on the subjects of limited war, NATO, aircraft carriers, and budgeting.[73]

President Eisenhower praised Sides's "dispassionate, low-key, and thorough exposition," but he criticized the report for failing to "identify those weapons systems which may be obsolescent, antithetical or overlapping." Arguing for the need to find a balance between the health of the economy and defense expenditures in the federal budget, Eisenhower told the NSC that "unless tough decisions were taken regarding such systems, in view of the fiscal situation we [will] find that in the long run we [will] encounter increasing difficulty in preserving our way of life if we put unnecessary money and resources into the machinery of war." The president directed the JCS to provide him with a study in which it evaluated weapons systems with this in mind. "Then . . . some very tough decisions [will] have to be made," he said.[74]

The endorsement, in WSEG-30, of a target plan similar to the alternative undertakings or minimum deterrence infuriated Air Force leaders. This type of deterrence would require a strategic nuclear force strong enough only to ensure that a retaliatory strike could destroy a large number of Soviet cities. Lieutenant General John K. Gerhart, the Air Force deputy chief of staff for plans and programs, immediately sent Gray a letter in which he criticized the target list of 135 Soviet cities. Gerhart worried that the White House might accept the list as "a valid and sufficient objective." Soviet military capabilities, he asserted, should be the primary focus for two reasons. First, counterforce targets represented the only opportunity to reduce damage to the United States in a general war.

> A concept of military action that disregards attack on vital military force is more than dangerous. . . . Looking first at the Soviet missile capability, we preserve U.S. and allied cities, people and military power by destroying at the very least his missile control and support capability, thus assuring that even if a first salvo has been launched, a second salvo will not be. . . . We realize our objective . . . by recognizing that the priority objective is military power. We cannot protect ourselves and our allies by any other concept.

Second, counterforce attacks by the United States would "include major destruction of population centers," but attacks on Soviet cities would not necessarily destroy military targets. Although he acknowledged that strikes against cities might enable the United States to achieve its wartime objectives,[75] Gerhart warned that "there is, I believe, still another danger in this concept of city targets. I fear its impact on the effectiveness of our deterrent. We must not gear our deterrent power to what would deter us if we were aggressively inclined but rather to what is most certain to deter the Soviet Union. . . . We cannot settle for a deterrent to which reasonable doubts must be attached."[76] Gerhart objected to WSEG-30 primarily because it failed to emphasize the necessity of damage limitation and that the Soviets might not be deterred by enormous civilian deaths.

Shortly after the briefing on WSEG-30, White House aides formulated a set of issues to be addressed in follow-up studies by the JCS. On October 30, Gray sent the list of issues to CJCS Twining, informing him that "the President attaches a great sense of urgency to this additional investigation." This list raised a number of questions about defense policy, including strategic nuclear forces. In regard to ballistic missiles, Gray sought the Chiefs' views on the balance between Titan, Minuteman, and Polaris, as well as between different bombers in the overall force posture.[77] These follow-up studies would not be completed until well into the next year—long after the administration had completed the defense budget for FY 1960.

In mid-November 1958, the NESC study on alternative retaliatory systems which Eisenhower had directed in March was briefed to the NSC. The report allayed Eisenhower's concern that it might be "too academic"; in Gray's opinion, it "could be an important springboard for proposing vital questions looking to the future of our defense posture." Specifically, Gray urged the president to order a Defense Department study that compared military targeting with mixed military-civilian targeting. This study could be linked to both budgeting and NSC 5904. Gray told the president: "Inasmuch as our military force requirements, including numbers and types of weapons and delivery systems [are] based entirely upon the purposes of our retaliatory force, the targeting system [is] central to our long-range planning with overwhelming implications for future defense budgets."[78] After hearing the report, President Eisenhower told the NSC that "he could remember well when the military used to have no more than 70 targets in the Soviet Union and believed that destruction of these 70 targets would be sufficient. Now, however, a great many more targets had been added."[79] He then approved Gray's proposal for a DOD targeting study.[80]

In January 1959, Gray met with the Chiefs, DOD civilians, and CJCS Twining to develop guidance for the NESC targeting study.[81] They decided the study should evaluate the "relative merits" of three different target systems: (1) primarily military, (2) urban-industrial, and (3) an "optimum mix" of the other two. Gray, who wanted a comprehensive report, asked McElroy also to address force posture.[82] Twining ordered an NESC comparison of the three target systems to determine the "minimum number of enemy targets" necessary to be destroyed by each system to "achieve the objective of prevailing in general war."[83] To facilitate an estimate of the "adequate" force complement for each target system, the study would employ the rates for weapons in-commission, their survival of surprise attack, reliability, enemy defense suppression, enemy resistance survival, delivery effectiveness, weapons per carrier, and re-strike availability.[84] The report, according to Twining, should also evaluate each target system with regard to the ability of forces to deter an attack delivered with both strategic and tactical warning.

Gordon Gray brought the debate over nuclear targeting and strategic force posture to the NSC and presidential decision-making levels through WSEG-30, the NESC study for 1959, and the JCS follow-up studies. It is clear that Eisenhower wanted to use these reports in deciding how best to achieve a stable strategic deterrent force and restrain defense spending. The Air Force found both discouraging. Its victories in the BNSP debates would prove to be bittersweet if targeting doctrine no longer emphasized counterforce, for such a change would mean a diversion of budgetary funds. Throughout 1959, the Air Force battled with the JCS to prevent this result.

After reviewing WSEG-30 and the guidance for NESC 1959, Defense Secretary McElroy sought the Chiefs' opinions on the "adequacy" and "overlap" of strategic weapons systems. The J-5 of the Joint Staff reported on January 14, 1959, that it had found "the over-all combination of weapon systems for strategic delivery to be adequate" to meet the Soviet ICBM threat, but not "inflexible." The J-5 based this conclusion on "the most measurable evaluation of the adequacy of our strategic delivery forces and weapons systems . . . is in the contribution of these forces and weapons to prevailing in general war."[85] But the Army and Navy continued their attack on counterforce targeting, rejecting the J-5's endorsement of the status quo in U.S. strategic weapons systems. An angry Air Force Chief of Staff Thomas White said the JCS split was "inconceivable" and accused the Chiefs of failing in "their corporate responsibility."[86] Ignoring the split, CJCS Twining notified Secretary McElroy on January 20 that "the programmed forces . . . are adequate" until such time as technology or international conditions "necessitate" adjustments.[87] But White remained incensed. The position of the Army and Navy, wrote the Air Force chief, "would not only give greatly increased stature to the 'Alternative Undertaking' targeting concept, but would in fact make the 'Alternative Undertaking' tantamount to the sole objective of nuclear offensive forces and the basis for establishing the nation's deterrent posture." White could not conceal his hatred for minimum deterrence: "I am convinced that the quickest way for the United States to lose its life and freedom would be to adopt the city strategy." The Air Force, he believed, should emphasize the necessity of damage-limiting retaliatory strikes but avoid making "rigid" distinctions between military targets and population targets. Destruction of Soviet military strength required attacking a variety of targets, ranging from nuclear delivery systems to the "governmental control structure."[88]

The debate resurfaced in June 1959 when the J-5 requested JCS guidance in evaluating the adequacy of strategic forces as part of the NESC study. A major difficulty was that "the principal purposes of U.S. deterrent forces are to contribute to the attainment of the two alternative basic national objectives concerning general war, viz., deterring general war or prevailing if general war occurs. It may be possible to have forces adequate to deter general war and yet not adequate to prevail if war occurred, despite deterrence, as a result of miscalculation. Conversely, a force may be adequate to prevail but may not be adequate to deter." Again the JCS split; the Army and Navy argued for a reduction in the number of strategic forces, since they believed the ideal target system was the "optimum mix." Both attacked the Air Force, advocating a reduction in the number of manned bombers either immediately or in the near future. Army Chief Taylor explained that "the appropriate target systems during the period under consideration should be an optimum mix of combined military-industrial targets and that the weapons systems for strategic delivery presently planned . . . are out of balance when compared to this threat, and in excess of those required, particularly in the numbers of [bombers]."[89] The Chiefs papered over these differences and informed McElroy that "Soviet technological advances will probably continue to diminish the margin of U.S. military superiority."[90] Still, the J-5 recognized the same aspect of strategic force planning that had occupied Paul Nitze over nineteen months earlier: employment strategy determined the optimum strategic nuclear force posture, but the two major strategies imposed different requirements for weapons deployments. Consequently, structuring nuclear forces in accordance with one strategy might preclude use of the other.

The intensity of the JCS debate over the adequate level of strategic forces resulted from the budgetary ramifications of increased force levels. Should the alternative undertakings and the population targeting strategy gain prominence, lower force levels would be adequate for deterrence and the Navy's invulnerable Polaris would become the strategic weapon of choice. The Air Force would lose control over targeting strategy and funding for second-generation counterforce weapons. The Air Force tried to undermine Polaris development numerous times. In June 1958, General Curtis LeMay advocated the "integration" of the command of all nuclear weapons, which would result in a de-emphasis of Polaris development.[91] When the other services challenged the adequacy of strategic forces in early 1959, the Air Force retaliated by demanding that it, rather than the Navy, should have operational command over the soon-to-be deployed Polaris. It argued that such command was justified because the submarine was inherently a strategic nuclear weapon.[92] General White proposed the formation of a "single unified command for strategic warfare" named "the United States Strategic Command" (USSC).[93] The new command would be divided into two sections: Polaris, and all Air Force strategic nuclear weapons. This reorganization, White maintained, would enable the United States to organize its strategic deterrent in the most efficient manner. White also proposed the dismantling of SAC with establishment of the USSC; he was confident that Air Force and SAC officials would dominate the new organization. But CNO Burke immediately rejected the proposal, claiming that conditions did not justify the formation of such a command because Polaris development could be coordinated through existing JCS mechanisms. Burke stated that Polaris would add "no imposing target coordination problems," for "the Polaris system will be targeted against the industrial base and the governmental control structure of the enemy—a relatively stable target system which readily lends itself to preplanning."[94] Burke endorsed the "disestablishment" of SAC, perhaps preceded by the devolution of SAC operational responsibility to the unified commands.

Unable to resolve the dispute over Polaris operational command, the JCS forwarded a split paper to McElroy in early May. Burke claimed that the "divergences on the concept of command stem from basic differences of philosophy." The "effectiveness" of Polaris, he wrote, would be dramatically decreased if the submarine were separated from other naval vessels: "[Polaris] cannot, and never was intended to stand alone as a missile/submarine combination awaiting a directive from *any* authorized source to fire. *Intimate* to and inseparable from the system are the many facets of naval operations at sea such as communications, and the close integration and coordination with other naval forces."[95] The USSC, he vehemently asserted, is "militarily undesirable, fiscally extravagant and unsupportable by facts, analysis and by logic." Burke recommended that Polaris operations be placed under the unified commanders with the commander in chief, Atlantic, (CINCLANT) retaining responsibility for operational planning. After a delay due to the sudden death of Deputy Defense Secretary Quarles, McElroy decided against establishing a USSC.

Between the disagreements over nuclear targeting and those concerning operational control of Polaris, CJCS Twining had had enough. Because he felt that the split within the JCS had "hampered current planning," he asked the Chiefs to answer fourteen questions for Secretary McElroy. Among them were:

> What should be our policy for development of a national strategic target system?
>
> What categories of targets should be included in the national strategic target system?
>
> What agency should apply our strategic targeting policy, develop the national strategic target system, and keep it up-to-date?
>
> What agency should review the national strategic target system for consistency with policy and approve it as a basis for further analysis?
>
> Do we need a single integrated operational plan for attack of the national strategic target system?
>
> If we do need [an SIOP] for strategic attack, what agency should develop this plan? What agencies should review it and approve it? . . .
>
> Is there an immediate need for the establishment of a Unified Strategic Command? . . .
>
> If a Unified Strategic Command is not established in the proximate future, is there a requirement for the integration of operational plans for the employment of Polaris submarines with CINCSAC's operational plan?[96]

Twining's disgust pushed him toward centralized planning as a means to resolve the long-standing JCS splits over nuclear strategy. He decided that U.S. nuclear planning would require significant alterations if it was to meet future challenges. However, it would be a year before the problem would be resolved.

When the JCS splits continued into 1960, Gordon Gray and Eisenhower decided that a firm presidential decision on the NESC "optimum mix" study might break the logjam.[97] At a special meeting on February 12, 1960, Lieutenant General Thomas Hickey briefed the NSC on the results of the NESC study of military (counterforce), urban (population), and optimum mix targeting. The NESC study supported the optimum mix system because it had "no major limitations," whereas both counterforce and urban targeting had inadequacies. According to the study, the optimum mix should result "in the U.S. prevailing in a general war."[98] The NESC listed 2,021 targets for executing this strategy and reported that a 75 percent to 90 percent certainty of destruction per target would ensure deterrence through 1963.[99] President Eisenhower decided that the lower limit should guide force planning as a way to inhibit nuclear stockpile expansion. Several days later, Gray recorded that he

> point[ed] out to [Eisenhower] that the difference between the 75 percent and 90 percent assurance of deliverability made a tremendous difference in numbers of weapons required and in force levels required. I therefore wanted him to be fully conscious of what he was doing when he approved the 75 percent figure. I reminded him that he had withheld final approval of a portion of a Defense recommendation with respect to the atomic stockpile pending the outcome of the study. . . . I pointed out that the Hickey presentation indicated that for the 75 percent capability our forces and stockpile were now adequate, whereas for the 90 percent capability they were not adequate, and my purpose was to make sure that at some point attention was directed to leveling off in stockpile requirements.[100]

But Eisenhower stood firm regarding his choice. Chairman of the JCS Twining described the NESC study as "a tremendous step forward" and thought it should be "a 'point of departure' for JCS planning." The JCS now had a clear presidential directive on nuclear targeting and the level of nuclear adequacy.[101]

President Eisenhower's decision on the optimum mix signaled his growing disenchantment with the overkill that characterized U.S. nuclear strategy. Eisenhower had received numerous briefings on the scale of potential casualties in a nuclear war, and each time he had registered his disgust and horror. But prior to 1959, these feelings had not caused him to change his positions on nuclear targeting or weapons procurement. The development of ICBMs armed with thermonuclear weapons pushed casualty projections to even more absurd levels. In mid-1959, Eisenhower received a top secret report stating that a nuclear war would kill 400 million people if the superpowers detonated a total of 10,000 MT (7,000 MT by the United States and 3,000 MT by the Soviet Union). No part of the world would be spared; there would be 80 million deaths in the United States, 140 million in Western Europe, 80 million in the U.S.S.R., 70 million in Asian communist states, and 30 million in Eastern Europe. A map of the Eurasian land mass showed a blanket of deadly radioactive fallout spreading across Europe and Asia that would cause inestimable additional deaths and inflict genetic damage for generations to come.[102] As far as can be determined, these figures exceed all previous estimates of nuclear war effects. In light of this report, Eisenhower's decision, months later, to limit the amount of overkill takes on new meaning. For the first time since taking office, he had made a major nuclear strategy decision on the basis of the level of destruction rather for than fiscal or strategic reasons.

Despite President Eisenhower's approval of the NESC targeting study, the armed services remained split over implementation of the targeting plan and operational control of Polaris. The president had already indicated to CJCS Twining that "Polaris must be drawn in to the over-all plan and tightly coordinated."[103] Defense Secretary Thomas Gates held a series of meetings with the Joint Chiefs in the late spring of 1960 but was unable to resolve their differences. In June, Air Force Chief of Staff White moved to end the logjam by proposing that, rather than create a new command, they give SAC responsibility for nuclear targeting.[104] After fifteen JCS meetings and additional Air Force briefings on the issue,[105] on July 6 Gates met with President Eisenhower to discuss a course of action. The defense secretary called the existing mechanism for nuclear targeting "cumbersome and expensive in nuclear raw materials" and claimed that the "management is bad." He also contrasted the approaches of the Navy and the Air Force to nuclear planning: "With some weapons the Navy does not plan to deliver warheads on target for some fifteen days following initiation of hostilities. On the other hand, the Air Force has developed a highly integrated set-up within the Strategic Air Command for targeting. They have spent much time on this subject and have the resources to spend on elaborate computers, etc."

Rather than creating a USSC, Gates recommended that instead, SAC be given responsibility for developing an "Integrated Operational Plan" and should use the Hickey NESC report as the framework for the first plan. "The only unit capable" of performing the function, he asserted, "is SAC. . . . Unless [SAC] is employed as an agent of the JCS it will not be possible to achieve a coordination any better than a tidying up of what we now have." President Eisenhower, complaining of Navy parochialism, agreed with Gates's recommendations.[106]

Secretary Gates formulated a draft directive according to which the SAC commander would also serve as the Director of Strategic Target Planning and in that capacity would issue a National Strategic Target List. The director would oversee the Joint Strategic Target Planning Staff (JSTPS) who would develop the Single Integrated Operational Plan for nuclear strikes against targets on the list. Gates circulated the draft to the JCS in August but delayed issuing the directive until after CNO Burke had made his case directly to the president.[107] On August 11, President Eisenhower met with Gates, Deputy Defense Secretary James Douglas, CJCS Twining, and CNO Burke at the defense secretary's request to attempt to resolve the dispute. After outlining his draft directive, Gates explained that a SIOP "will work and that it will put the JCS effectively in supervision over SAC where this has not existed previously." Burke then stated his argument against this "radical" proposal: "I am fearful that if the responsibility and the authority for making a national target list and for making a single operational plan is delegated to a single commander—for all commanders forces, the JCS will have lost control over operations at the beginning of a general war, even though they will be given the opportunity to review and approve the national target list and the operational plan." The changes, he expected, would adversely affect the conduct of wartime operations by unified commanders and undermine United States relations with NATO member nations. Burke pleaded with the president to give targeting authority to a strengthened Joint Staff rather than to the SAC—or at least to obtain the views of the unified commanders before approving Gates's directive.

President Eisenhower concurred with Burke regarding the possible weakening of NATO and the desirability of strengthening the Joint Staff, but he rejected his conclusion that CINCSAC should not have responsibility for targeting decisions. Such steps toward integration were necessary, he said, since "the initial operations of the future impose a requirement for greater rigidity in planning than in past methods." General Twining, in a remarkable display of antipathy, claimed that the Navy would deliberately undermine the targeting process if the president implemented the directive on a "trial basis," as Burke suggested. The Navy, he told the meeting, "[will] succeed, as they have succeeded in obstructing any really effective coordination of target planning over the past ten years." Later he added: "[Polaris] is a purely strategic weapon, and we must do something about it to integrate it with other attack means. . . . [T]he crux of the problem is as it has been for nearly twenty years, that the Navy is completely opposed to serving under a single commander." Deputy Secretary Douglas reiterated the need for a new method of coordinating target planning; he reported that "there are as many as 200–300 targets that are subject to attack by duplicating systems." President Eisenhower politely rejected Twining's assertions of Navy parochialism but clearly and firmly stated his desire to establish an integrated plan: "For the first strike, there must be rigid planning, and it must be obeyed to the letter. After the initial strike, increased flexibility will be needed and should be incorporated in the plan." The president approved Secretary Gates's draft directive but reserved the right to make adjustments after he had reviewed the first SIOP—scheduled to be completed by mid-December.[108]

Eisenhower's decision constituted his second step toward formulation of an operational nuclear strategy. In February, he had decided that an optimum mix system with a 75 percent assurance of destruction should guide targeting and strategic force planning. This would limit the amount of overkill and decrease the Air Force's justification for expanded strategic forces to match the Soviets. Now, he had created the mechanisms for translating that guidance into an operational nuclear war plan. In taking these two steps, he tried to limit organizational influence by rejecting the services' preferred strategies. This integrated approach could be considered successful only if the first SIOP achieved these goals.

Throughout the fall of 1960, President Eisenhower made an extra effort to encourage oversight of CINCSAC Power's development of the first SIOP. Both in meetings of the NSC and in private meetings, Eisenhower said "he wants the Joint Chiefs to keep right in the middle of the SIOP." His concern emanated from his "considerable doubts as to General Power's capacity in this duty." The new Joint Chiefs chairman, Lyman Lemnitzer, assured him that the JCS and the Navy members of the JSTPS understood the president's desires and that Power knew "that the activity will be closely observed."[109]

By October, the SAC-dominated JSTPS had produced the United States' first integrated nuclear war plan. The massive strike called for in the SIOP would wreak havoc on the Soviet Union and result in innumerable casualties.[110] The plan translated the optimum-mix system into 1,060 "designated ground zeros" (DGZ) with the ratio of military targets to civilian targets of about 4 to 1—emphasizing the counterforce side of the mix. Depending on the amount of warning, the size of the U.S. striking force ranged from 1,004 delivery vehicles to the total alert force. Just prior to the plan's effective date in 1961, the SIOP total alert force consisted of 1,108 weapons (1,023 aircraft weapons, 30 cruise missiles, and 55 ballistic missiles) with a cumulative yield of 1,798 MT. The SIOP total force on the effective date consisted of 3,012 weapons (2,808 aircraft weapons, 130 cruise missiles, and 74 ballistic missiles) having a total yield of 7,420 MT.[111] The yield of this striking force exceeded that projected in the 1959 estimate of the effects of fallout that had been reported to Eisenhower. Overall, the plan called for massive strikes against the entire Sino-Soviet bloc and allowed for little flexibility or discrimination. When CJCS Lemnitzer briefed President John F. Kennedy on this plan, he stated that "the SIOP is *designed* for execution as a whole. . . . [I]t must be clearly understood that any decision to execute only a portion of the entire plan would involve acceptance of certain grave risks."[112]

In early November, the president sent science adviser George Kistiakowsky to Omaha to review the SIOP.[113] Kistiakowsky, who was greatly disturbed by the plan's assumptions and options, said he believed it "will have major repercussions."[114] On November 25, Kistiakowsky and a PSAC staff member, George Rathjens, briefed President Eisenhower on the SIOP. Kistiakowsky explained that the assumption used by the JSTPS had produced "misleading results." The president "strongly agreed that the function of judgment concerning the over-all operational objectives should not be turned over to experts and planners." In his opinion, Polaris should be used only in a second round of nuclear strikes—thus reducing the level of "over-kill." "This type of planning," the president told the two men, "fails to recognize that war of the kind described no longer makes any sense." On the basis of the SIOP's damage criteria, the science adviser said, eight large nuclear weapons would have to be targeted against a city the size of Hiroshima in order to produce the same amount of damage as inflicted by the first atomic bomb. For this reason, Kistiakowsky advised Eisenhower to "[approve] this plan subject to a thorough and complete reworking. This is a first trial run and is a good piece of work, but it should not become immutable in the next administration. There should be a thorough revision, including the basic thinking such as attack in all one phase rather than in successive waves." Aware of the gravity of the issue, Kistiakowsky handed Goodpaster his papers after the November 25 meeting and told him that "questions of this kind raised in [them] should somehow be put in the hands on the next administration."[115] President Eisenhower, despite his disappointment with the SIOP, took no formal action which enabled the Joint Chiefs to adopt it on December 2.

During these months, Air Force headquarters organized a plan for reasserting a purely counterforce strategy. The Air Force's continued organizational dominance in targeting had been recognized in the SIOP; nevertheless, Eisenhower's decisions on optimum mix and fiscal policy restricted the development of new weapons systems. The Air Force provided a justification for counterforce—and for the weapons to support it—by projecting an increase in Soviet targets into the 1970s. The Air Force Directorate of Intelligence estimated that Soviet target lists would reach enormous proportions, increasing from 1,450 DGZ in 1960 to 3,800 in 1970. Since each DGZ could encompass several targets, the Air Force anticipated that the maximum target list would increase from 3,560 in 1960 to 6,955 in 1970. According to the Air Force, the "failure to maintain the counterforce capability greatly alters the present power relationship in favor of the Soviet Union."[116] To prevent any further erosion of counterforce capability, the Air Force created the "New Approach Group," which would, in effect, "sell" counterforce ideas developed by the Rand Corporation to the incoming Kennedy administration. The Air Force thought it could regain primacy for a counterforce strategy by attempting to influence Kennedy's national security transition team (headed by Senator Stuart Symington) and by placing Air Force officers in key positions in the office of the new Secretary of Defense.[117] These efforts had some success in the early 1960s when Robert McNamara had a brief fling with counterforce. But his reasons for rejecting it can also be traced to the events of these months—specifically, to WSEG-50.

In late December, the WSEG completed a study, begun sixteen months earlier, in which it examined the two main issues of nuclear strategy: targeting and strategic force posture. WSEG-50—including its eight volumes of appendixes—constitutes one of the most significant documents in the making of U.S. nuclear policy.[118] Although it never officially reached any members of the Eisenhower administration, the report is indicative of the intellectual ferment that had been taking place. It was a significant influence on the strategic thinking of some members of the Kennedy administration, Robert McNamara in particular.[119] Most important, the report provided continuity in nuclear strategy between the two administrations, despite their public disagreements over the missile gap.

The WSEG assessed the "relative effectiveness" of all strategic weapons systems planned for the mid-1960s on the basis of a range of criteria.[120] The cost per target destroyed served as the principal criterion for comparing the utilities of the different weapons systems—which varied, of course, with the chosen strategy. The special attributes of ballistic missiles, particularly "rapid delivery time and high confidence of successful penetration," made them the optimum weapon for striking the anticipated target system, according to the WSEG.

WSEG-50 included an extensive analysis of counterforce targeting. A first strike in which the Soviets detonated 3,000 MT, the WSEG established, would inflict casualties amounting to 90 percent of the U.S. population irrespective of the target system. If the United States responded to a Soviet attack with an immediate counterforce strike, a casualty rate of at least 80 percent of the U.S. population could be expected. Should the United States employ a counterforce *first strike* against the Soviet Union, the WSEG estimated that the Soviet Union would still deliver a 1,000 MT response—producing casualties ranging from 70 percent to 80 percent of the U.S. population.[121] When the WSEG studied the relationship between counterforce and Soviet force posture, it found that an increase in U.S. strategic forces would have rapidly diminishing returns, since an increase beyond slightly more than 1,000 weapons could not reduce the size of the Soviet retaliatory strike.[122] The estimated minimum percentage of the U.S. population that would be killed increased from 47 percent in 1963 to 63 percent in 1967, *regardless of the magnitude of the U.S. attack.* Adding to these frightening results, the WSEG warned that "the outcome of any realistic case is likely to be a good deal worse."[123] The group "thus concluded that counterforce alone does not appear to be a high confidence measure for preventing unacceptable levels of damage to the U.S. in the event of war." But minimum deterrence offered no greater appeal to the WSEG because it "would not only be ineffective in deterring overseas aggression, but might cause Soviet leaders to doubt that such a force would in fact be used in reply to their initial strike against our forces."[124]

WSEG-50 projected a reduction in civilian exposure to a Soviet nuclear attack (as did an earlier study of active defenses, WSEG-45) through the use of both active and passive defenses. Defenses could be of value to the United States regardless of the Soviet nuclear force posture. According to WSEG calculations, defenses could decrease casualties by 40 million to 60 million. But neither active defenses nor passive defenses for industry could prevent major damage. More important, "the cost of implementing these measures will probably be in the tens of billions of dollars. . . . [B]y employing technologically feasible countermeasures, e.g., cluster warheads, and increasing its missile force, the U.S.S.R. can maintain a strong retaliatory posture capable of doing great damage to the U.S. notwithstanding the implementation of defense measures."[125] It would be cheaper for the attacker to build more offensive weapons than for the defender to maintain active defensive systems.

The WSEG prepared a list of actions intended to strengthen both deterrence and the ability to wage nuclear war. For continued deterrence, it recommended that the United States maintain a force posture capable of inflicting "high levels of damage" in retaliation to a Soviet attack. To ensure the credibility of nuclear deterrence, the WSEG suggested limiting extended deterrence: "The most important effect of the nuclear stalemate upon our total posture is that it will curtail drastically, and perhaps eliminate, our ability to project U.S. strategic power, as now defined, into foreign areas in support of American diplomatic policies which are not immediately and directly crucial to our continued national existence."[126] Almost reluctantly, the group reported that improved counterforce targeting for missiles, defenses, and command and control, as well as possession of reserve strategic forces, would increase U.S. prospects for survival in a nuclear conflict. But the offense's inherent advantage made the situation discouraging: "Because present technology has provided the offensive with a variety of measures at far less cost and earlier than their defensive countermeasures, there can be little optimism in achieving these results." WSEG-50 concluded with another warning:

> Though a nuclear stalemate seems to be approaching and likely to remain for a considerable period, it must not be conceived as a static stalemate. Technology is progressing too rapidly to believe that the stalemate cannot be broken. All promising avenues of research which might break the stalemate to our advantage should be pursued vigorously. The nation that can indeed develop, for example, an effective active missile defense even in the face of countermeasures will be well on the way to achieving strategic superiority.[127]

Privately, WSEG members were less hesitant about stating their opinions on counterforce and the ideal strategic force posture for the United States. According to Rathjens, the WSEG found that "a counterforce U.S. policy does not make any sense from a purely military point of view unless there is a reasonable probability that it will be successful to the point of reducing the number of Soviet retaliatory weapons well below the 100 level."[128] But as the report showed, it would be very difficult to limit U.S. civilian casualties by employing counterforce measures.

In many important respects, WSEG-50 reflects the evolution of strategic force planning in the last years of the Eisenhower presidency; it also raises some of the unresolved questions that remained for the Kennedy administration. At the most basic level, the report reaffirmed the value of developing redundant strategic weapons systems. "Systemic redundancy," which both civilians and military leaders accepted during this period, would strengthen deterrence by ensuring execution of a retaliatory strike regardless of future Soviet weapons development or use of countermeasures. Possession of a triad consisting of bombers, SLBMs, and ICBMs would vastly complicate any Soviet planning for a first strike against the United States. The triad became a venerated concept in U.S. strategic thought and was often invoked to justify the initiation or continuation of a particular weapons system. But it should be remembered that the triad's intellectual roots lie in the very real search for a secure second-strike capability against immediate and future threats, both real and imagined.

WSEG-50 included a number of influential observations on the structuring of strategic forces (even though the WSEG members apparently withheld their true opinions). Foremost, it determined that the United States failed to accrue any strategic advantage beyond a certain level of nuclear weapons deployment. Recognition by the WSEG of the problems of a damage-limiting strategy was an important step in addressing the question of how much is enough. WSEG-50 also maintained (as had WSEG-45) that an active defense would be only marginally useful for the foreseeable future, since offensive countermeasures would be increasingly less expensive for the United States and the Soviet Union. The efficacy and cost-effectiveness of offensive forces in comparison with defensive forces limited the alternatives available in making nuclear strategy. The formulation of nuclear strategy in the missile age vexed the WSEG just as it did the Eisenhower administration. Both the WSEG (in this report) and the administration rejected the Navy's strategy of minimum deterrence. Studies conducted in the late 1950s, including WSEG-50, revealed the magnitude of destruction to both superpowers that would result from a strategy of counterforce targeting. The WSEG tempered its endorsement of counterforce strategy by recommending a decrease in international commitments and an emphasis on passive defense.

The Eisenhower administration's efforts to formulate nuclear strategy during the missile gap period resulted in the development of many ideas that are central to nuclear policy, including what has become known as mutual assured destruction. It tried to steer a middle course between minimum deterrence and counterforce targeting, but it ended up with a strategy that conformed, in general, with the latter. The logic for formation of a strategic nuclear triad was developed in WSEG-50 and other reports that advocated a force posture containing weapons with different characteristics. The reality of Soviet ICBMs—with potentially large numbers in the future—made the successful execution of a counterforce strike during a general war difficult, if not impossible. The alternative strategies designed for use against Soviet cities or against some mix of military and urban targets only ensured that a general war with the Soviet Union would end in cataclysmic destruction for the United States. Neither of the nuclear strategies available to the United States constituted an acceptable approach for the conduct of nuclear war. It was also becoming increasingly evident, with the growing perception of U.S. vulnerability, that the United States could not rely on extended nuclear deterrence.

The prospect of Soviet missiles and mutual vulnerability precipitated a major debate over national strategy and nuclear strategy within the Eisenhower administration during the missile gap period (1957 to 1961). Scholars have largely overlooked these policy areas because no new doctrine was enunciated and because the relevant documents remained classified for decades. The Eisenhower administration's policy reviews during this debate facilitated the transition from massive retaliation to mutual assured destruction. An examination of the debate is thus important if we are to understand the evolution of U.S. nuclear strategy.

The availability of recently declassified documents has enabled scholars to gain many important insights into the administration's decision-making processes as it struggled to keep pace with a changing external environment. The administration retained massive retaliation as the core of its national strategy for fiscal and strategic reasons despite continuing attacks on the policy within the administration. Defining a nuclear strategy to support the national strategy proved far more problematic. With respect to targeting, the administration settled on a hybrid of counterforce and minimum deterrence that, when implemented in the SIOP, resembled counterforce nonetheless. The administration also tried to establish the adequate size and composition of strategic nuclear forces. Decisions on nuclear strategy dragged out much longer than those on national strategy because of interservice rivalry and civilian dependence on military expertise. Consequently, the most important decisions on nuclear strategy came late in Eisenhower's second administration—allowing the services to resurrect the same issues under the next administration.

The Eisenhower administration's decision-making processes for national strategy and nuclear strategy differed despite the obvious relationship between the two areas. National strategy formulation was primarily an NSC exercise. As such, government institutions represented on the NSC participated in meetings of either the council or the NSC Planning Board. The making of national strategy depended only minimally on the expertise of the national security bureaucracy. This arrangement allowed for much bureaucratic politicking but not much obstructionist filibustering by recalcitrant organizations. In the making of nuclear strategy, the NSC's dependence on military expertise and evaluations reduced the council's control over the decision-making process. Unlike the national strategy debates, the deliberations outside the NSC by such groups as the JCS and the WSEG impacted on the NSC's capacity to make nuclear strategy. Reliance on military evaluations skewed the decision process by creating "action-channels" in which fewer individuals participated. The importance of military evaluations gave the armed services the opportunity to undermine or obstruct policy development, as was evident in the 1959 JCS debates. The making of nuclear strategy had more organizational content than did national strategy. The president tried to lessen organizational influence (but without complete success) by constantly prodding, maintaining oversight, and demanding specific studies.

The differences in policy making in each area reflected, to some degree, differences in the nature of decisions. Formulating national strategy required determining how to balance the nation's means and objectives—a judgment that could be made by only the president. But decisions about targeting doctrine or the force posture necessary for deterrence required various technical judgments, which altered the structure of the decision process. The president and other civilian leaders had to wait for the military before making such decisions. That the decision processes of two policy areas so closely related could differ so markedly demonstrates the need to exercise caution in evaluating the effectiveness of alternative decision-making models or different presidential styles.

Finally, this chapter presents insights into President Eisenhower's views on nuclear strategy which scholars have often held to be contradictory. He kept massive retaliation as part of U.S. national strategy for both strategic and fiscal reasons—he worried about the effects that an increase in conventional forces would have on the economy. Additionally, Eisenhower believed that any U.S.-Soviet conflict would escalate to a general war in which neither side would surrender. Any movement away from massive retaliation might weaken strategic deterrence and actually precipitate Soviet aggression. Viewed in this light, Eisenhower's ambiguous statements about the use of nuclear weapons during crises take on a different meaning. The president was not embracing nuclear warfare or being coyly deceptive but may have been signalling what he thought would happen and how the United States would *have* to act. But Eisenhower's determination with regard to national strategy undermined his position as to nuclear strategy. In refusing to abandon massive retaliation, the president kept a strategy that provided no guidance concerning the optimal targeting doctrine or force posture. In the absence of clear presidential directives, organizational interests produced interservice struggles in all important decisions about nuclear strategy. By the last year of his presidency, Eisenhower seemed to embrace many of the elements of MAD, due in large part to the casualty and damage estimates he had received. But with the last administration statement on national strategy completed in July 1959, these views were never fully articulated and the quandary of nuclear strategy was left for the next administration. As departing Secretary of State Christian Herter wrote on his list of problems that required the Kennedy administration's attention: "The nuclear balance, and the inevitable abandonment by the U.S. of massive reliance on massive retaliation. . . . The painful withdrawal symptoms as we move away from 'massive retaliation' to dependence on more useable types of military power."[129]

[4]

Defense Budgets and the Public Debate over the Missile Gap

Presidents occupy two roles in the policy-making process. First, they must be effective *executives* and mobilize the government bureaucracy into formulating and implementing policies that meet their objectives. Second, they must be effective *politicians* and gain support from outside the executive branch. Narrowly, they must secure congressional approval of their policies through the legislative process. Broadly, presidents need the public and elites to accept these policies as legitimate. Consequently, the policy arena demands that presidents be both executives and politicians. Performing these two roles effectively sometimes requires different leadership styles and practices. Presidents often gravitate to the role in which they feel more comfortable or adept. Ronald Reagan is a perfect example of a president who virtually ignored executive responsibilities in favor of political leadership. The two roles are rarely perfectly balanced, in part because of the nature of policy issues, the power of the executive branch, and the federal structure of government. Further, the lack of a clear distinction between the roles necessitates that presidents be competent in both. A president who consistently fails in one role is certain to encounter obstacles in the other. Circumstances arise when they must navigate troublesome shoals of the role in which they are less comfortable. This is perhaps the most difficult test of a political leader's skill and mettle.

The formulation of U.S. national security policy during the Cold War emphasized the president's executive skills due to the extensive secrecy of the issue and the national consensus supporting containment. Traditionally, the most publicly debated aspect of national security policy has been defense budgeting. Congress's exercise of its constitutional responsibilities have often provided the basis for public debates over the conduct of national security policy. This was especially true during the missile gap period. Congressional budget deliberations prompted a national debate over intelligence estimates, strategic weapons procurement, defense spending, and national objectives. The missile gap debate tested Eisenhower's leadership skills to their fullest extent.

President Eisenhower's major goal in defense budgeting was to maintain military capabilities sufficient to deter the Soviet Union without undermining economic health. He often reminded aides that the United States could easily deter the Soviet Union if it increased defense spending astronomically, but the imposition of additional taxes or other controls would ruin the free-market economy which might result in the establishment of a "garrison state." Thus Eisenhower wanted defense policy to walk a thin line that avoided both provocative underspending and ruinous overspending. This approach to defense budgeting emerged early in the first administration, with the New Look, and continued until January 1961 with little change. It was manifested through stable defense budgets which avoided any significant tax increases. Nuclear weapons served as the means for upholding the nation's defense commitments at low costs. Eisenhower consistently rejected any attempt to increase limited-war forces, on the grounds that a limited war with the Soviets would automatically escalate to a general conflict. This belief governed not only budget decisions but decisions concerning national strategy, as discussed in Chapter 3. But after Sputnik, Eisenhower's defense budget goals came under attack by elements within and outside the administration. The armed services demanded more funds for both conventional and nuclear weapons. Congressional Democrats, journalists, and academic strategists challenged Eisenhower's budgetary limits, his massive retaliation strategy, and the validity of intelligence estimates. Guiding defense budgets through this tumultuous environment required all of Eisenhower's executive and political leadership skills.

The public debate over defense budgeting and the existence of a missile gap provides an opportunity to assess the strengths and weaknesses of Eisenhower's leadership style. Many revisionists, particularly Fred Greenstein, have praised Eisenhower's leadership style and have virtually ignored its weaknesses. But the genius of Eisenhower's leadership style may not be as clear as these revisionist scholars assert.

The declassified documentary record reveals that Eisenhower's leadership frequently produced success and failure in the same policy area. Eisenhower succeeded in his executive leadership by formulating defense budgets that conformed to his objectives. However, his political leadership is markedly mixed. While the Congress passed his budgets without significant amendment, Eisenhower never articulated a case for his policies which would quiet the missile gap critics permanently. This failure of political leadership is due to both the nature of the policy issue and Eisenhower's preferences in leadership styles.

FY 1959 Defense Budgeting: Spring to October 1957

The public debate over defense spending during the missile gap period began in the spring of 1957. The Eisenhower administration's BNSP review for 1957 raised fears that defense spending would soon have a destructive effect on the economy unless remedial actions were adopted in the next budget. This fear emanated from several factors. First, the New Look weapons programs had entered the development and deployment phases almost simultaneously, producing an enormous demand for dollars. To maintain a stable defense budget and fund the new programs, the administration would have to find cuts in some other areas or raise taxes. Second, Eisenhower's economic advisers cautioned that the economy could not support a large defense budget for economic and political reasons. The Council of Economic Advisers estimated that four years of economic growth would not necessarily obviate the need for deficit spending. But Treasury Secretary George Humphrey believed that continued deficit spending would send the economy into a deep recession. He pointed out that mounting public pressure for a tax cut only exacerbated the situation: "Short of a greatly increased and more obvious threat of war, it [is] going to be almost impossible for the Administration to avoid a tax reduction over the next couple years. Domestic politics make this almost certain." Humphrey recommended that, to prevent deficit spending, expenditures should be cut when taxes were cut. The president admitted: "Certainly we in this Administration have done our utmost to keep expenditures down. We now feel a certain sense of defeat." Only John Foster Dulles maintained that higher taxes would not constrain economic growth, but he failed to sway other administration members.[1]

Economic pressures on the Eisenhower defense program accumulated rapidly throughout the spring of 1957. The first indications of an impending recession began to appear.[2] In April, an FCDA report recommended that the administration initiate a shelter program at a cost of $32 billion over the next decade.[3] Yet just as DOD expenditures began to exceed authorized funding levels the House of Representatives cut $2.5 billion from the administration's defense budget.[4] President Eisenhower grew increasingly impatient and seemed unusually testy whenever the economics of defense spending were discussed. For example, when Humphrey noted the pressure for a tax cut, Eisenhower asked curtly, "How much [would you] be willing to pay to save [your] life?" He also complained that "everyone seemed to be 'going nuts' " on research and development programs.[5] In an effort to avoid escalating defense expenditures or weakening defenses, the president directed Defense Secretary Charles Wilson to plan on a $39 billion ceiling for the next five defense budgets. This would necessitate cuts in current spending (FY 1957), in the budget then before Congress (FY 1958), and in the budget the Defense Department had just begun to formulate (FY 1959).[6]

In mid-summer, Wilson showed Eisenhower his budget plans for FY 1959. As the president had directed, expenditures and new authorizations would be kept below $39 billion. This level would be difficult to maintain, however, because Congress authorized less for FY 1957 than the administration had requested ($37 billion), whereas the Defense Department expended funds at a rate of $40.25 billion for the year. Unobligated funds from the previous year could cover the immediate shortfall. But inflation ensured that the $38 billion defense budget planned by Wilson would purchase even less defense than it had in the previous year. Wilson recommended reducing the number of Defense Department personnel from 2.8 million to 2.5 million over the next two years, cutting military construction, restricting funds for research and development, and limiting weapons procurement, especially of missiles. He explained that "the magnitude of the [missile] program as originally contemplated and its increasing cost due both to technical difficulties and basic cost increases is just too big."[7] These cuts, he said, represented "an absolute minimum in the absence of real improvement in the international situation." Eisenhower thought the budget plan was "generally good" approved it for presentation to the NSC.[8]

The changes in the military budget were approved without much input from the armed services. The JCS effectively excluded themselves from the debate by presenting CJCS Radford with a five-year plan that exceeded the budgetary ceiling by several billion dollars annually and even ran as high as $43 billion.[9] Radford told Wilson coolly, "I feel little purpose would be served to ask them [the JCS] to rework their submissions," and offered even deeper reductions (a $37.3 billion defense budget and 600,000 fewer troops). He justified these deep cuts on the grounds that new weapons provided greater firepower—essentially a further extension of the New Look.[10] Radford's position eased any administration concerns about facing a military united in opposition to the cuts or facing charges that it ignored military advice in developing the program.

With Eisenhower, Wilson, and Radford in agreement, the defense budget guidelines were presented to the NSC. At a July meeting, Wilson emphasized to the NSC the need for personnel reductions and the redeployment of troops from Europe. Inflation, he noted, had added 5 percent or $2 billion to the defense budget in the previous year: "U.S. forces must be reduced if Defense [is] to live within its $38 billion ceiling, but . . . continued inflation might make it impossible to maintain this ceiling." Wilson and Radford commented that their authority had increased because of JCS disagreements; Radford added that he had "repeatedly emphasized to the Joint Chiefs of Staff the fact that their failure to agree automatically required that their decisions be made for them by higher authority."[11] Both he and Deputy Defense Secretary Quarles warned that this defense program couldn't respond to every change in Soviet capabilities and might be difficult to defend to Congress. But the president weighed in, explaining that Congress had already cut defense expenditures more than he believed was adequate, so even deeper cuts could still occur. He told the NSC that "he *expected anyone who had anything to do with the Defense budget to defend the figure of $38 billion a year.*"[12] The incoming Joint Chiefs chairman, General Nathan Twining, endorsed the plan but noted that it would require a "substantial redeployment." Treasury Secretary Humphrey repeated his belief that the economy could not continue to support such high defense expenditures. Eisenhower agreed, but commented that only "a safeguarded disarmament" would enable the deep cuts Humphrey sought. Wilson said he thought the American public would support a $38 billion defense budget, especially once economic growth had decreased the share of government expenditures earmarked for defense. The key concepts for the FY 1959 defense budget were established, with the president demanding loyalty from his administration for the final preparations.[13]

Wilson concentrated on missile programs in his effort to reduce defense expenditures below Eisenhower's ceiling. Missile programs seemed ripe for paring because of the number of programs, their high costs, and low immediate military return. According to the DOD, missile development through 1957 had already cost the U.S. $11.8 billion, but in the next six fiscal years the figure would escalate to $36.1 billion. The NSC was, according to Eisenhower, "thunderstruck" when it found this out.[14] Wilson estimated only 10 percent of the budget, or $3.8 billion, could be allocated each year for missile development. Without adjusting for inflation, that meant an average cut of $2.5 billion per year. Wilson proposed several ways to reduce missile expenditures: elimination of overtime spent on IRBMs (except for flight-testing); appointment of a committee to consolidate Army and Air Force IRBM programs; elimination or reduction of several short-range missiles; and a limitation on funding for the Titan ICBM. The president and the NSC approved Wilson's missile cuts in late July without significant opposition or debate.[15]

By the end of summer 1957, the president and his senior national security advisers had settled on a defense program for the next five fiscal years. The anticipation of high budgets, deficit spending, and ruinous inflation prompted the administration to make reductions in force size, weapons development, and overall spending. Under this plan, the defense budget would gradually shrink as inflation weakened the government's purchasing power. This plan emanated from a decision process concentrated in the NSC and in private meetings between Eisenhower, Wilson, Radford, and Humphrey. The armed services played an inconsequential role in setting the budget and personnel ceilings. Defense Department civilians and CJCS Radford discounted JCS advice because the Chiefs ignored budgetary ceilings in making their recommendations. In a reversal of the traditional bureaucratic decision-making models, the *civilians* presented the armed services with a fait accompli; they set the limits for the budget and specified the allocations to each service. The civilians brought the services back into the decision process only to settle some questions about final inter-service allocations. But the launch of a 184-

pound metal sphere in early October upset the basic budgetary assumptions.

Sputnik, the Public Uproar, and the FY 1959 Defense Budget

Sputnik I shocked both the American public and the Eisenhower administration. It demonstrated Soviet mastery over rocket technology beyond that of the United States. Americans linked this technological feat to presumed hostile Soviet intentions leading to a precarious view of U.S. security. Pro-defense Democrats, academic defense analysts, and media elites found the public unusually receptive to attacks on the Eisenhower administration. The administration's initial efforts at reassuring the public of the military insignificance of Sputnik I only backfired, making it seem even more out of touch.[16] Newspaper editorials castigated the administration and cited the summer missile cuts as further evidence that it was ignoring the missile race.[17] This created a dicey political situation for the administration. Reassuring the public of the adequacy of U. S. defenses while limiting defense spending would be a tough sell after Sputnik. Moreover, the economy had been in a recession since August, and the political dynamics within the administration had changed. The three officials who had been instrumental in setting defense priorities (Wilson, Radford, and Humphrey) had all departed the administration by early October. Their replacements encountered treacherous political waters upon taking office.

As public pressure built after Sputnik I, the Eisenhower administration put the finishing touches on the defense budget for FY 1959. Initially, it agreed that defense expenditures of approximately $38.6 billion would be sufficient for that year.[18] But events in early November forced the administration to reconsider its decision. First, the Soviets launched Sputnik II on November 3. This second Soviet satellite weighed 1,121 pounds and had a life-support system for its canine passenger. These two feats further dramatized how far Soviet rocketry skills had advanced beyond the United States.[19] The day after the launch of Sputnik II, the Gaither panel delivered a second shock to President Eisenhower, in a private briefing. The Gaither report recommended actions to improve strategic retaliatory power, warning, and defenses that would require expenditures totaling $44 billion over the next five years.[20] The panel asserted that the nation could afford this program if it increased both taxes and the debt limit, reduced expenditures in other areas, and took steps to curb inflation. According to the Gaither panel, defense spending as a percentage of GNP should increase from 10 percent to approximately 13 percent. When the panel met with the NSC on November 7, Eisenhower questioned whether Congress and the public would accept the constraints increased defense spending would impose on the economy. "What . . . can the American people be expected to put up with in terms of the allocation of the Gross National Product over the next several years? Was the Panel proposing to impose controls on the U.S. economy now? . . . Are we now to advocate the re-introduction of controls?"[21] Unilaterally and contrary to the administration's instruction, the Gaither panel expanded the scope of its study and developed an itemized program rather than the "broad-brush" study desired by the President.[22] Still, the administration could not ignore the report because of the tenor of the times; it became the benchmark for future programmatic reviews.

The two Sputniks, public criticism, and the Gaither report caused the president to renege on the $38 billion defense ceiling.[23] Defense increases would have to be made cautiously so that they would not justify greater deficit spending to counteract the recession. Several days after being briefed on the Gaither report, Eisenhower told the new defense secretary, Neil McElroy, that defense expenditures could be increased as high as $39.5 billion if the additional funds went to urgent programs such as SAC alert and pay increases to retain skilled personnel.[24] Eisenhower and McElroy agreed that the NSC would review the $38.6 billion defense budget while the services developed a series of supplemental budget requests.

When McElroy presented the unamended defense budget to the NSC in mid-November, Eisenhower and others acknowledged that high defense spending would have economic effects but agreed that the existing situation necessitated a spending increase. The JCS complained that the result of a $38 billion defense budget "would be a reduction in military capabilities from the existing posture."[25] McElroy reviewed the service requests for $4.87 billion in FY 1959 supplemental spending.[26] After consulting with the JCS, the service secretaries, and the president, he reduced the original $38.6 billion budget but added $2.14 billion, to reach a defense budget totaling $39.5 billion. The NSC approved increased funding for strategic weapons, expanded bomber alert and dispersal, and several other programs. As the meeting concluded, Eisenhower told the NSC that "when the Council had first become involved directly in the ballistic missiles programs [I] had expressed the opinion that the effect of ballistic missiles would be more important in the psychological area than in the area of military weapons. [I] still [feel] that as a weapon the manned bomber was superior to the missile. We [are] now, however, in a transition period but it [is] still a question [of] how much money we are justified in asking for on behalf of these missile programs."[27]

A $38 billion FY 1959 budget had become the first casualty of the missile age. Eisenhower rationalized the increased defense spending by thinking that "greater confidence in security" might lift the economy out of the recession.[28] By approving the additional budget increment, administration officials grudgingly admitted that a critical period had begun. But the president never lost sight of his objective of stable defense expenditures and tempered budget accelerations accordingly. This approach to defense spending would become a source of political controversy throughout the remainder of his presidency.

President Eisenhower recovered from the initial Sputnik maelstrom and responded in an effort to reassure the public.[29] A nationally broadcast speech on November 7th concerning science and national security constituted the centerpiece of his efforts. The President announced the creation of the position of DOD director of guided missiles and of the Office of the Special Assistant to the President for Science and Technology (OSST), with the appointment of James R. Killian as the first presidential science adviser. This institutional restructuring reflected Eisenhower's dissatisfaction with management of the missile program over the past two years (see Chapter 5). But these measures seemed too restrained, given the volatile climate. In the Senate, majority leader Lyndon Johnson (D-Tex.) held explosive hearings on the U.S. and Soviet missile programs, in the hope of riding Sputnik to even greater national recognition. The committee's extensive witness list ensured that administration critics would have ample opportunity to air their charges. In early December, the president intensified his lobbying by unveiling the defense program for FY 1959 to the congressional leaders. Eisenhower and his aides claimed that the new defense budget would adequately meet the increasing Soviet threat, particularly with the recent supplemental items. Congressional leaders seemed impressed by Eisenhower's presentation, offering little criticism or commentary.[30]

Eisenhower's success in mollifying his critics soon evaporated when word of the Gaither report spread outside the administration and around Washington. Bitter because of the administration's cold response to the report, Gaither panelists conducted a campaign within the government promoting their recommendations since November.[31] For example, Nitze accused John Foster Dulles of sabotaging the panel's passive defense recommendations and demanded that the secretary resign to end the nation's "crisis of confidence."[32] Eventually, the *Washington Post* published the gist of the Gaither report on December 20th.[33] Suspicious that Eisenhower wanted to suppress the report to prevent increases in defense spending, Democrats called for the release of a sanitized version (as some Gaither members and Vice-President Nixon privately agreed). After several weeks of such pressure, Eisenhower considered releasing a summary of the Gaither report but ultimately decided against doing so.[34] Some panel members agreed with the administration's decision but others encouraged continued pressure try to to force the administration to take action on the panel's recommendations.[35]

President Eisenhower's cold response to the Gaither report and his decision not to release a sanitized version should not be interpreted as a wholesale rejection of it, as some authors have argued.[36] In fact, the NSC scheduled four meetings in the winter of 1957 and the spring of 1958 to examine twelve Gaither panel recommendations in detail—most of them focusing on strategic nuclear forces. But any expansion in force levels would require yet another budgetary increase even before FY 1959 began.[37] At the first meetings, the NSC planned to review the Defense Department's evaluation of three Gaither panel recommendations on missiles: ICBM and SLBM force size; missile hardening; and active defense. But organizational rivalry engulfed the JCS and effectively prevented an expedited review even though the administration seemed willing to loosen its purse strings further. This split allowed recommendations from outside the JCS to set the debate.[38]

Unable to secure agreement at the JCS level, the services forwarded individual requests that together required an additional $10 billion for strategic programs in the next two years. McElroy, benefitting from the independent technical evaluations of James Killian and his staff, reduced the service proposals to $1.648 billion.[39] He recommended $400 million for two new Polaris submarines, $100 million for Titan improvements, $100 million for solid-fuel technology research and development, $225 million for Nike-Zeus, $456 million to produce five B-52 bombers per month in the next year, and $91 million for development of a new air-to-surface missile (ASM) to improve B-52 performance.[40] At 3:00 in the afternoon of March 20, Killian met with President Eisenhower to review McElroy's proposed budget increases. The president agreed with many of the recommendations, especially the two relating to the B-52. An hour later, they were joined by McElroy, Quarles, Twining, Budget Bureau Director Maurice Stans, and White House staffer Jerry Persons to discuss the recommendations.[41] After discussing the program, Stans warned of the fiscal impact of the proposals; he pointed out that the FY 1959 military budget would exceed $42 billion after this second post-Sputnik expansion. More important, Stans noted, the commitment to these research programs also meant a commitment to increased budgets in the future. Eisenhower responded that the services must request a realistic program—not accelerations amounting to $10 billion. He told the group that "hopefully" future military budgets would not exceed $40 billion.[42]

The political climate created by the Gaither and Sputnik shocks ensured a smooth path through the Congress for the proposed defense increases. Congressional debate of defense issues in the spring of 1958 focused on the Defense Reorganization Act rather than the budget. Legislative leaders in both parties seemed satisfied with the President's initial responses to the Sputniks and the criticism abated. But some in the administration other than Stans were growing uncomfortable with the economic implications of the defense program, in part because of tax cut proposals designed to ease the recession. When the NSC discussed the missile program in April, the president warned of "unheard-of inflation" if strategic weapons development proceeded without "some very hard thinking" about the point at which weapons become obsolete. John Foster Dulles challenged the ultimate objectives of the military program: "The United States should not attempt to be the greatest military power in the world, although most discussions in the Council seemed to suggest that we should have the most and best of everything. . . . In the field of military capabilities enough was enough. If we didn't realize this fact, the time would come when all our national production would be centered on our military establishment." "Too much [defense]," Eisenhower reiterated, "could reduce the United States to being a garrison state or ruin the free economy of the nation."[43] No one in the administration, at least at this time, appreciated that any attempt to reassert controls over defense spending would be politically difficult.

The FY 1960 Defense Budget: July to December 1958

Economic concerns strongly influenced the administration's initial FY 1960 defense budget decisions. Eisenhower and his senior advisers sought greater NSC centralization of the budget process to counteract pressure from the armed services for higher defense spending. While Eisenhower and CJCS Twining pressed the JCS to make economically sound budget decisions, McElroy informed them that he would make the budget decisions if they could not.[44] Assistant Secretary of State Gerard Smith and several NSC staffers wanted a formal mechanism for the review of defense budget priorities by the NSC Planning Board.[45] Defense Department budget evaluations conducted in the summer of 1958 justified the economic concerns. They estimated annual increases in defense spending of $1 billion annually over the next four years. Without new taxes, the United States would incur the largest peacetime budget deficit and have the highest national debt in its history. Steadily increasing personnel and weapons costs were eroding the DOD's purchasing power. Under these circumstances, warned an assistant defense secretary, "we could not expect. . . . to have a satisfactory rate of modernization or to undertake any desirable new programs."[46] This prediction jeopardized Eisenhower's hopes that the FY 1959 defense budget was a one-time aberration and that subsequent defense budgets could be kept below $40 billion.

In October 1958 the armed services submitted a draft FY 1960 budget. They requested expenditures totaling $43.1 billion, an increase of more than $2 billion over the total for the previous fiscal year. Nearly $20 billion would be spent on weapons procurement. They also proposed a whopping $7.3 billion increase in budget authorization from $41.1 billion to $48.4 billion. The size of this budget submission reflected the services' continuing inability to agree on the division of limited dollars. The draft budget, with its allocation for expensive technologically advanced weapons, only intensified economic concerns, for the administration saw its control over defense spending slipping away. Stans told McElroy that the "estimated projections of these programs to subsequent years indicate that a number of very difficult defense program decisions must be considered if the serious fiscal situation which faces the government is to be minimized."[47]

The staffs of Stans and Killian went to work on the draft and each formulated possible budget reductions. In separate memoranda, they emphasized the need to balance defense expenditures with available funds by reducing or eliminating various weapons programs.[48] The Bureau of the Budget advocated measures to bring the defense budget back down to the $40 billion figure desired by Eisenhower. Most significantly, the BOB believed that authorizations should not exceed expenditures in the next fiscal year; otherwise the administration would be locked in to some type of increase in future years. Attainment of this objective would require "very restrictive policies for support and procurement programs." Killian's staff developed a list of technical questions that, they hoped would force the Defense Department to address the question of "how much military capability do we need and how much more or less will it cost?" Both the BOB and the OSST advocated reducing the number of strategic systems in development. The latter focused on the size and composition of the strategic force posture on the grounds that the administration still had not satisfactorily addressed this issue.[49] The BOB and the OSST also questioned the funding for a variety of weapons, including Nike-Zeus, Polaris, Titan, Nike-Hercules, Regulus II, the B-58 and B-70 bombers, the KC-135 tanker aircraft, the Thor and Jupiter IRBMs, the Dynasoar, and a new air-launched ballistic missile (ALBM) for use with the B-52.[50]

As the final budget decisions approached, John Foster Dulles endorsed cutting strategic programs rather than conventional forces. He told Treasury Secretary Robert Anderson that "I felt that we needed at least our present conventional weapons establishment, but I thought that we could cut down on the nuclear effort on the theory that all we needed there was enough to deter; that we did not need to be superior at every point." Dulles also informed Gordon Gray that he "had a feeling that 'massive retaliation' was somewhat overdone with overlapping projects, and that the danger now was that there would be more than needed there and not enough" in conventional forces. Since Dulles would not be attending the defense budget conferences, he asked Gray to convey his desire "that budget decisions would not impair or cripple our capacity to deal with local situations."[51]

In late November, McElroy circulated a draft FY 1960 defense budget that included expenditures of $41.6 billion and authorizations of $42.8 billion—amounts substantially lower than the service requests but still about $1 billion more than in FY 1959. The budget called for significant increases in strategic forces (but no crash programs). The number of Polaris submarines would increase from 6 to 12; programmed ICBM deployments would increase from 130 to 200 (90 Atlas and 110 Titan), B-52 production would continue at the rate of 5 aircraft per month but the deployment of B-58 medium bombers would decrease. The Budget Bureau criticized McElroy's draft for continuing an inexorable movement toward a $45 billion budget within one or two more years. It advocated achieving a stable budget through further reductions in weapons procurement and at least another 100,000 fewer personnel.[52]

Eisenhower met with McElroy and other senior administration officials on November 28 to review the defense budget before presenting it to the NSC. As McElroy outlined the budget figures and major cuts, Eisenhower admonished the group: "We have wasted too much money by going prematurely into production. . . . It is not a question of *either* defense *or* fiscal solvency. You do not have defense without fiscal soundness." McElroy defended the duplication between ICBM programs and argued for continuing Titan, on the basis of advice from "the scientific community." In justifying the B-52 and B-58 expenditures, he cited B-47 structural fatigue and Soviet active defenses. Dismayed and exasperated, President Eisenhower asked "why we produce B-58s along with more B-52s. He [concurred] with the policy of introducing a few new high performance aircraft for psychological purposes, but he [questioned] the retention of Titan, Atlas, B-52s and B-58s. He voiced the question, 'How many times do we have to destroy Russia?' "[53] When McElroy had finished his presentation, BOB Director Stans challenged the budget on fiscal grounds. He "congratulated" McElroy on the reductions but asserted that expenditures had to be reduced to $40 billion to avoid a budget deficit. Having been provided by his staff with a list of possible cuts totaling $8 billion, Stans argued for an even stricter evaluation of the defense budget to eliminate "exotic" weapons such as the nuclear-powered aircraft, especially since the Budget Bureau had cut the nonmilitary part of the federal budget by an enormous $5.5 billion (from $42.0 billion to $36.5 billion).[54] Stans informed the group that "as the President's budget officer, he could not recommend approval of the budget as presented by the Department. His recommendation was to instruct the Defense Department to utilize the figure of $40.0 billion as a ceiling in a fresh attempt to make up a new FY 1960 budget." Eisenhower did not comment on this recommendation but reminded his aides that if defense spending was so high that the budget was not balanced, then taxes would have to be raised. He emphasized forcefully that "he wants everyone in the Department to have a complete understanding of the total problem of the interrelationship between national defense and the balanced budget, [and] *that unless the budget is balanced sooner or later, procurement of defense systems will avail nothing*."[55] The president directed the Defense Department to "reexamine" the draft budget before presenting it to the NSC.

The conflict between the goals of increasing strategic forces to meet the Soviet threat and maintaining low expenditures to achieve a stronger economy dominated the NSC defense budget review on December 8. Encouraged by the president's attacks on defense spending and overlapping weapons systems, Stans and Treasury Secretary Anderson came prepared to cut the defense budget even more.[56] To minimize service dissent at the meeting, McElroy had secured the JCS' tepid support several days earlier.[57] With increased defense spending out of the question, the Defense Department emphasized the budget's risks and tried to prevent additional cuts. This budget, McElroy noted, would "dislocate employment and trade. . . . One or more aircraft corporations . . . might have to go out of business." Similarly, CJCS Twining and DOD Comptroller Max Lehrer explained that funding for the procurement of major conventional weapons (ships and aircraft) did not even meet the level of replacing those lost to attrition—thus posing the risks of obsolete weapons or fewer forces. Anderson lectured the NSC on the dire effects of high defense spending, budget deficits, and increased taxes. "We have to get into the problem of whether this country can invariably afford every right gun and every right target at every right time. . . . When our military people look at all these weapons systems," he concluded, "They must see what other things we are trying to defend and where money is being spent in this country. We must try to protect the American competitive system." Stans proposed that the NSC reject the draft budget until further cuts could be agreed upon, adding that "we might need the help of the President" to accomplish this. Eisenhower snapped back, do you think I've been sitting on my hands?

Throughout the meeting, President Eisenhower seemed troubled by the overlap between missile systems and proposals for their expansion. He complained that "we [are] putting too much money on Atlas" (which remained technically unproven) and questioned the requested increase in Polaris submarines from 6 to 12. "You [can] not win if you [persist] in putting your money on all the colors of the wheel." McElroy pointed out that an increase in strategic weapons was needed to maintain the nuclear balance with the Soviets. But after CNO Burke outlined Polaris developments, Eisenhower asked: "How many times do we have to calculate to destroy the Soviet Union? These Polaris missiles . . . ought to have some regulatory effect on" ICBM programs. He said he agreed with "the need for a good little fleet of Polaris submarines, but do we need such a hell of a lot of them?" Sensing that support for the budget was unraveling, McElroy attempted to justify his program: "The budget did not contain a gun for every need. There *were* calculated risks in this budget. We [are] *not* moving as fast as we could with proved weapons systems. In the matter of long-range missiles, it . . . constituted a relatively conservative program. . . . Many hard long sessions with the best people available to us made this budget the best that the President's Defense team could present." McElroy received some support from Killian, who claimed that the budget included "calculated risks" in the areas of air defense and limited war forces, although some savings might still be possible. At the end of the long meeting, Eisenhower said he approved the budget "in general" but reminded the group that final preparations should ensure that "all practicable economies [are] effected."[58]

The principals left the meeting with differing interpretations of Eisenhower's directive. Stans, Gray, and the president himself believed that Stans and McElroy had been ordered to meet and develop further reductions. McElroy felt the NSC meeting represented the end of the budget process, although he continued to meet with Anderson to discuss the budget's effect on the economy. Gray informed Eisenhower of the differing interpretations and added that "McElroy [is] not prepared to make any meaningful adjustment" to the defense budget. An irate president replied that "if [I was] in charge, [I feel I] could take $5 billion out of the Defense budget." Gray attempted to calm the president, explaining that reductions of that magnitude could be achieved only by cutting weapons programs. Eisenhower concurred, and said reluctantly that "if Defense, after all of the meetings and conversations on the subject, still [maintains] that the programs presented in the NSC meeting [are] essential to the national security, [I have] little choice but to approve them." Moments later, Eisenhower phoned McElroy to express his dissatisfaction with the budget process. Gray recorded that:

> the President indicated that he had been "dragooned" into approving the Defense programs as presented. He made it clear to Mr. McElroy that his

> approval was reluctant but it was given only because he felt he had no choice. He continued to have, however, reservations about the numbers of Atlas and Titan missiles, wondering if it was necessary to program as many of both. He then said that he wanted Mr. McElroy to get together with Mr. Stans right away and subject to further discussion the non-programmed items which he had mentioned to me earlier.[59]

Shortly thereafter, McElroy reduced FY 1960 defense expenditures to $40.95 billion; by making a number of minor revisions, he was able to limit the increase in defense expenditures to $145 million.

Formulating the defense budget for FY 1960 frustrated Eisenhower greatly, for it had become increasingly difficult to maintain both strategic superiority and tight fiscal controls. The New Look formula of simply transferring funds from conventional forces to nuclear weapons was no longer practical because of the high cost of advanced weapons and because multiple systems were being developed simultaneously. Reluctant to impose an additional burden on the economy, Eisenhower constrained defense spending by cutting the amount spent for conventional weapons procurement, reducing the number of personnel, and restricting the increase in strategic weapons systems. This compromise resulted in a defense budget that fulfilled Eisenhower's objectives but exposed the administration to attacks by advocates of both "crash" programs for missile production and expanded limited-war forces. Eisenhower's decision to limit defense spending for both conventional forces and nuclear weapons systems pleased neither group and gave them common ground to attack the administration's program.

The Rise of the Missile Gap Debate

The first sign of a renewed political debate over the nation's defense programs came in the summer of 1958. Some congressional Democrats believed that the acceleration in the U.S. missile program made in early 1958 fell short of the crash program that was required in light of intelligence reports on Soviet missiles.[60] The result might be a "critical period" during which the Soviet ICBMs force far exceeded that of United States.[61] These concerns, as well as rumors of even more dire intelligence data led to Senator Symington's inquiry into the validity of intelligence estimates (discussed in Chapter 2).

By December 1958, the administration failed to convince Symington of the validity of the NIEs, despite its best efforts. Symington was now confronted with the question of whether he should keep his dissent over missile estimates private. He decided to go public with his charges for three reasons. First, he had been a strong advocate of air power, while serving both in the executive branch and in Congress. For Symington to endorse a force posture that constrained air power would be tantamount to a renunciation of a decade of work. Second, Symington's plans for a candidacy for the 1960 Democratic nomination for president were becoming increasingly obvious. He hoped to use the defense issue to achieve greater national recognition. Third, Symington worked closely with the Convair Corporation, the prime contractor for Atlas—the ICBM that was then nearest deployment. Symington had been told of the intelligence rumors that prompted his assertions of CIA underestimates of Soviet missile development by Thomas Lanphier, a former Symington aide and vice-president of Convair. Convair stood to gain or lose from any alteration in the projections of Soviet missile capabilities. If the Soviet ICBM threat became more acute and the United States needed to deploy ICBMs quickly, it could turn only to Convair's Atlas. Any downgrading of the Soviet ICBM threat might allow the United States to delay ICBM deployment or even defer major deployments until the second-generation Minuteman (not produced by Convair) was available. Symington therefore prepared a public challenge to the administration's strategic forces program and intelligence estimates of Soviet missiles while offering an Atlas crash program as the solution.

The opening of the 86th Congress gave Symington and other critics an ideal public forum. Bolstered by the largest congressional majorities since the New Deal, the Democratic leadership immediately engaged the administration after it unveiled the defense budget for FY 1960 at a confidential briefing on January 5, 1959.[62] Senator Lyndon Johnson told the press that he was "deeply concerned and somewhat disappointed to observe that in the field of military preparations they [the administration] are programming as if we were living in a static world rather than an exploding, expanding and developing world."[63] Symington and other Democrats criticized the administration for not proposing higher defense spending even though the intelligence estimates predicted that the Soviets would have a missile advantage in several years. The Democrats received support from academic circles as well. *Foreign Affairs* published Albert Wohlstetter's influential article which warned of a precarious nuclear "balance of terror."[64] President Eisenhower's attempts to reassure the public seemed only to confirm the Democratic charges. He openly acknowledged the existence of a Soviet missile lead yet cautioned Americans not to "disturb ourselves too much that we have not yet caught up with another great power and people with great technical skill in a particular item."[65]

Senior administration officials defended the administration's programs and intelligence estimates in both closed and open congressional hearings held in January 1959. McElroy and Twining gave a remarkably candid summary of NIE 11-4-58, the current BNSP (NSC 5810/1), and U.S. military capabilities in testimony before the Senate Foreign Relations Committee meeting in executive session. (Twining later repeated the briefing for the Senate Armed Services Committee.) The briefing made no effort to downplay the Soviet threat. Twining described the Soviet threat as "implacable hostility" that had an "aggressive . . . program of world domination." His accompanying charts graphically depicted such images such as that of a Sino-Soviet spiked boot crushing the globe.[66] He warned the senators that NIE projections "represent only an estimate of what the Soviets could produce and deploy," not what they would actually do.[67] Twining admitted that U.S. forces had deficiencies but argued against making numerical comparisons of military capabilities; he claimed that "our nuclear retaliatory forces continue to provide the United States with a margin of advantage which, if exploited effectively in conjunction with other military operations, would permit the United States and its allies to prevail in general war."

Some committee members responded favorably to the presentation and expressed support for the administration. One senator who remained unconvinced , however, was John F. Kennedy (D-Mass.). He questioned Twining extensively concerning the validity of predictions of a missile gap, prompting the following response from the exasperated Joint Chiefs chairman:

> My point, as I said a minute ago, is let's don't pick one weapons phase in isolation and call it a gap.
>
> For instance, we will have IRBM's in Europe—consider what we have to attack Russia with—that are better than ICBM's. Look at the 15 bombers sitting on the border, and the Matador missiles and the Navy equipment.
>
> *We are surrounding them.* The only thing they can hit us with is the ICBM in the missile field, and we can hit them with all kinds of missiles.

> *Don't you see the difference?* Compare these things collectively. Ours is a collective defense; it is not United States solely. We would like to beat them on ICBM numbers; maybe we won't. But that doesn't mean we have lost the war.[68]

After the hearing, some Democratic committee members, including Kennedy, publicly criticized the administration for failing to acknowledge the gravity of the situation.

Such differences in focus were typical of the missile gap debate on Capitol Hill in the first months of 1959. No matter how hard administration officials tried to persuade Congress of the adequacy of the defense program, some Democrats focused instead on the ICBM gap predicted in the intelligence estimates. Press reports also highlighted the missile gap aspect of the defense debate. After McElroy told a congressional committee that the United States would not race the Soviets "missile for missile," a front-page story in The *New York Times* began: "The Secretary of Defense testified today that the United States was voluntarily withdrawing from competition with the Soviet Union in the production of intercontinental ballistic missiles."[69] The erroneous perception spread that the administration had ceded ICBM superiority to the Soviets. The congressional hearings soon revealed the existence of a split between the administration and the armed services over the budget. McElroy disclosed the December 1958 JCS certification of the budget's "adequacy" but admitted that each service had "reservations" about the budget.[70]

This revelation opened the door for the armed services to attempt end runs around the administration to seek congressional approval of a budget increased. The armed services and their congressional allies coordinated their efforts to change the defense program. W. Barton Leach, a Harvard Law School professor and longtime consultant to the Air Force chief of staff, advised Kennedy to "keep hammering" away at the administration's policies in order to "gradually make certain policies [budget ceilings] disreputable and thus deter their continuance."[71] Symington publicly claimed that the United States could deploy many more ICBMs in the next few years than the administration had planned—clearly an allusion to Convair's Atlas. Privately, Symington pressed Twining to purchase more Atlas missiles, but to no avail.[72] These tactics infuriated both Eisenhower and Twining. The president called Symington "neurotic" and a "demagogue" who "leaked security information" that "misled" the public. He told Killian and some PSAC members that "the munitions makers are making tremendous efforts toward getting more contracts and in fact seem to be exerting undue influence over the Senators." The president identified Symington by name in case anyone failed to make the connection.[73]

Congressional hearings held in the late spring gave each service an opportunity to articulate publicly their differences with the administration's budget. Air Force Chief of Staff White supported the budget, though he admitted it contained weaknesses, especially in the area of bomber procurement. Army Chief of Staff Taylor openly criticized the budget's ceiling and personnel reductions, which, he believed, would weaken U.S. limited war capability. Chief of Naval Operations Burke complained about the decrease in the number of ships. Congressmen in both parties openly attacked Eisenhower's competence in military matters. Symington made wildly inflated predictions about Soviet missile strength: "In three years the Russians will prove to us that they have 3,000 ICBM's. Let that be on the record."[74] The congressional testimony by top officials of the armed services "greatly disturbed" Eisenhower because it raised issues previously decided by the administration. He complained to the NSC that the integrity of the JCS had been diminished. Since Congress would exploit such divisions for "partisan advantage," Eisenhower "insisted" that "every military man should support the final decision of those in positions of authority after he has had the opportunity to state his own personal view. . . . Suppose . . . we were actually in a state of war and all these differences of opinion and challenges to authority were being aired?"[75] White House aides suggested that the president travel around the country and make some speeches to try to regain the political momentum on defense, but he made only a few.[76]

The missile gap issue declined as the defense budget made its way through Congress. Although many congressmen chided the president about defense priorities, few were willing to increase taxes or to make dramatic changes to match their rhetoric.[77] Congress reshuffled the funds allocated to limited war forces and air defense—resulting in appropriations that were $19.9 million *less* than the administration's request.[78] In a narrow sense, congressional reluctance to make changes represented an important budgetary victory for Eisenhower despite the growing political hostility to his program. But the defense debate damaged Eisenhower politically in three important ways. First, it revealed administration disagreements about the scope and direction of the defense program. Second, it promoted an image of the president as detached and unconcerned about Soviet missile progress. Third, it allowed Democratic senators to capitalize on the missile gap issue and to present themselves as patriotic defenders of national security.

The Defense Budget for Fiscal Year 1961: Spring to December 1958

As the congressional debate over the missile gap issue and defense spending progressed in early 1959, Army Chief of Staff Maxwell Taylor noticed a change in the political milieu. For the first time, the Eisenhower administration requested a JCS budget certification of approval as a form of political protection from congressional critics.[79] Recognizing an opportunity to increase JCS leverage over defense budgeting, Taylor proposed that the JCS accompany its Joint Strategic Objectives Plan with a statement of the forces required for five missions: nuclear retaliation, conventional forces, reserve forces, air defense, and naval forces. Joint Chiefs Chairman Twining thought a JCS budget review might improve the corporate image of the JCS within the administration by demonstrating that the armed services agreed on most spending issues. He approved a procedure for budget review in which each service would propose funding for all the armed services over the next three years and within a ceiling that assumed a 5 percent annual growth rate. The JCS would combine these proposals, resolve disagreements over specific programs, and use the final product to formulate the defense budget for FY 1961.[80]

Not surprisingly, the services' budget projections reflected the parochial interests of each. The Army recommended shifting funds from strategic weapons—which it considered already "adequate"—to limited war forces and active defenses (both Army programs). The Navy proposed additional funds for conventional weapons procurement and for a limited expansion of strategic forces that emphasized "dispersal and mobility"—in other words, Polaris submarines over SAC missiles. The Air Force, which was already receiving about half of the defense dollars, proposed maintaining the status quo. Contrary to Twining's expectations, the services agreed on only 38 percent of the budget items (24 of 63). The most controversial issue was the allocation of procurement funds for strategic programs such as Nike-Zeus, Titan, Minuteman, Polaris, and strategic bombers. Although these funds constituted only 10 percent of the defense budget ($4 billion to $5 billion), each service fought vigorously for them because their distribution would affect future mission capabilities and access to defense budgets.

JCS budget coordination deteriorated in the summer of 1959 after McElroy set a ceiling of $42.2 billion for the FY 1961 defense budget (including construction). The services formulated their budgets independently and delivered them to McElroy during the first week of September. In submitting its budget, the Air Force complained that McElroy's budget ceiling provided $1.2 billion less for FY 1961 than would be needed to maintain its program unchanged. The Air Force requested $19.3 billion—$500 million more than McElroy's ceiling for it—which, it said, would require cuts in personnel and in weapons procurement programs, resulting in "substantial risk" to the nation.[81] The other services shared the Air Force's concern over continued reductions in weapons procurement and personnel. McElroy reviewed the budget requests and made numerous cuts amounting to several billion dollars. The JCS conducted another budget review but was hopelessly deadlocked because of interservice rivalry. As McElroy prepared his defense budget presentation for Eisenhower, it became clear to him that Taylor's initiative aimed at increasing JCS leverage had failed. Interservice competition for shrinking funds had stymied the process. The JCS deadlock left the difficult decisions about defense priorities in the hands of DOD civilians and, ultimately, President Eisenhower.

JCS complaints about the defense budget began to register with McElroy and Eisenhower, who were becoming less comfortable with their economic constraints. McElroy told the president that the JCS doubted that the United States "will have what it really requires" for defense. Aside from cutting the authorized funding level for ships and constraining missile development, maintaining stable defense spending would require further reductions in forces and a decrease in foreign commitments—primarily that to NATO. Eisenhower responded, "[I] had not realized the budgetary situation is this severe." He said that McElroy's report "disturbed" him. Several days later, Eisenhower took McElroy aside after an NSC meeting and told him that the defense budget could exceed the ceiling, even if such an increase meant cutting nondefense spending. Encouraged by the attempts of the JCS to limit spending, the president directed McElroy to be "flexible" in making budget decisions. McElroy said he "appreciated" the advice and promised not to abuse it.[82]

Eisenhower's willingness to loosen the purse strings helped McElroy only marginally. The demands for funds for different missions (nuclear retaliation vs. conventional forces) and weapons systems within the missions could be reconciled with fiscal restraint only by making hard choices about the direction of U.S. defense policy. The administration's BNSP papers ensured the primacy of nuclear missions. But within the strategic missions, a variety of missiles and bombers competed for development funds. With many weapons entering the most expensive phases of development and deployment, the nation could not afford them all. The most important decision the administration confronted was the role of strategic bombers in the missile age—a decision that only the president had the authority to make.

The key budget decisions were made in a series of meetings held in November in Augusta, Georgia, where Eisenhower was vacationing prior to his three-week trip to Europe and the Middle East. On November 16, McElroy and Twining discussed the remaining budget decisions with Eisenhower and his staff. After contemplating reducing NATO commitments and reserves, Eisenhower criticized the defense program for "trying to defend ourselves against every conceivable type of weapon" rather than just responding to Soviet programs. McElroy began to outline an "off-the-shelf" airborne alert for 25 percent of the B-52s, but the president interrupted: "It looks again as though we are trying to protect ourselves in several ways at once." Despite his obvious dissatisfaction, Eisenhower admitted that he "did not have the time or the skill or the knowledge to set up arbitrary" budget decisions. He rejected a proposal by the Budget Bureau that it conduct its own defense budget review. McElroy's draft budget increased programmed ICBM forces from 200 (90 Atlas and 110 Titan) to 270 (130 Atlas and 140 Titan), although Titan deployments would be delayed until storable liquid fuel became available. Eisenhower endorsed the Atlas force levels but delayed a decision on the Titan force levels. (He approved McElroy's recommendation in January 1960.) He also approved the additional procurement of three Polaris submarines and long-lead-time funding for procurement of another three.

McElroy then introduced the subject of bomber programs, by far the most contentious nuclear weapons issue. The DOD proposed terminating funding for the high-speed B-58 medium-range bomber and limiting its deployment to 49 planes; the Air Force and Twining wanted another $1 billion to increase deployments to 90 planes. With very little discussion, the president approved the Air Force proposal, since the B-58 had already entered production. But he criticized the B-70, the successor to the B-52: "If we expended effort on the B-70 we would simply be saying that we had lost all faith in missiles." After extensive discussion of the B-70 (see Chapter 5), Eisenhower told McElroy to make the budget "leaner and tougher," but no agreement was reached on B-70 funding.[83]

Several days later, Eisenhower brought the Joint Chiefs of Staff (including the new Army chief of staff, Lyman Lemnitzer, who had replaced Taylor), to Augusta. The president wanted to ensure administration unity on the defense budget and thus prevent a repeat of the congressional defense debate earlier that year. In a long lecture, he criticized the European Allies, who "lean much too heavily on us." Referring to the economic burden of U.S. international commitments, the president complained: "We tend more and more to get other people into the habit of expecting us to pick up the responsibilities and the costs." Air Force Chief of Staff Thomas White made the case for additional B-70 funding—well aware that the future of manned strategic bombers hung in the balance. Twining and Lemnitzer approved of continuing the B-70, but at lower funding levels.[84] White's defense of the B-70 "impressed" Eisenhower but failed to persuade him to approve the Air Force program.[85]

President Eisenhower met with the service secretaries on November 21 to review the controversial budget items. He told them that he recognized the need for redundancy but rejected full development of the B-70 because it would become operational years after ICBMs. In Eisenhower's view, the B-70 "supplemented" ICBMs instead being a "transition" to them. His belief that a nuclear war in the missile age would target cities buttressed his position on the bomber. Under these conditions, the B-70 would add only marginally to strategic capabilities since its main advantages were controllability and accuracy in counterforce targeting.[86]

Meanwhile, McElroy put the finishing touches on the defense budget. He limited the budget increase to a mere $100 million and allocated only $74 million for development of the B-70.[87] The budget met Eisenhower's programmatic goals. The budget curtailed a number of weapons (particularly IRBMs and Nike-Zeus), expanded ICBMs and Polaris submarines, kept conventional forces low, and maintained fiscal ceilings. The major defense issues had been settled at Eisenhower's Augusta meetings, so the NSC quickly and uncontentiously ratified McElroy's budget on November 25. But the president's leadership style and his preparations for the upcoming budget battle with Congress were much in evidence at this NSC meeting. Despite the *pro forma* atmosphere of the meeting, Eisenhower questioned military officials extensively on specific spending levels, overlap of strategic programs, and the B-58 just as he had those attending the meetings in Augusta. For example, the president "asked whether the Air Force would not have at some time to reach the hard decision as to whether it wished to concentrate on planes or missiles." Repeating some of his Augusta questions at the NSC meeting served two purposes: it placed Eisenhower's concerns in the NSC official record; and it alerted the rest of the NSC of the grounds for the choices made in the budget. The president asked McElroy to cut another $200 million to give him "a big lever" for forcing more nondefense cuts: "Indeed, [I wish] to use a reduction in the Defense Budget as a club, not only against the departments and agencies, but against Congress."[88] With his trip abroad only days away, the president seemed satisfied with his programmatic and political preparations for his last defense budget battle with Congress.

The Second Missile Gap Debate

Just after President Eisenhower returned to the United States, his defense program was already under attack. Maxwell Taylor, who had been retired from his position as Army chief of staff for less than six months, published *The Uncertain Trumpet*. The cover displayed Taylor's four stars of rank and boldly announced:

> General Taylor contends:
>
> —that the doctrine of massive retaliation has endangered our national security;
>
> —that our military planning is frozen to the requirements of general war;
>
> —that weaknesses in the Joint Chiefs of Staff system have left the planning of our military strategy to civilian amateurs and the budget-makers.
>
> And General Taylor presents a new national military program to correct the resulting deficiencies.

Taylor proposed a "flexible response" strategy to meet a wider range of contingencies without having to resort to nuclear weapons. His reports of discussions by the JCS and the NSC presented an unflattering picture of the Eisenhower administration's decision-making process.[89] Others attacked Eisenhower's leadership in defense policy; Samuel Huntington argued that defense policy lacked direction because of a legislative-like decision process in the executive branch.[90] Furthermore, Senator Henry Jackson (D-Wash.) was busy preparing a congressional investigation of the NSC. The stage seemed set for another act in the drama over the missile gap.

On January 18, Eisenhower submitted a $79.8 billion budget to Congress and called for restrictions on spending so that surplus funds could be used for deficit reduction.[91] In congressional hearings held a few days later, Defense Secretary Gates and CJCS Twining attested to the defense program's adequacy. Responding to administration critics, Twining bluntly told the Senate Armed Services Committee: "There is no deterrent gap."[92] Gates revealed that the latest NIE (NIE 11-4-59—which was yet to be officially "published") on Soviet missiles was predicated on Soviet intentions not production capability (see Chapter 2). This resulted in lower estimates of Soviet missile deployments over the next three years: "We do not now believe that the Soviet superiority in ICBM's will be as great as that previously estimated." Senators left the presentation with generally favorable opinions.[93] Unfortunately, DCI Dulles had presented a much less optimistic estimate to the Senate Foreign Relations Committee only days earlier. He had detailed the Soviet missile test program, improvements in reliability, and the numerical projections in NIE 11-4-59. He told the senators that "we have not picked up any evidence on fixed ICBM bases, or for that matter movable bases," but that the Soviets could have as many as ten operational ICBMs.[94]

When Secretary of State Christian Herter appeared before the same committee in closed session on January 21, many committee members confessed their confusion after comparing the Dulles testimony with that of Gates and Twining. They demanded an explanation of the disparity and wanted to know why intentions, rather than production capability, now served as the basis for the NIEs.[95] Herter also expressed bewilderment and offered to clarify the incongruities at a later date.[96] In subsequent testimony, Gates amended his position, stating that the Soviets could have 150 ICBMs operational by mid-1961 and that they would then outnumber American ICBMs by a ratio of three to one.[97]

This stark projection prompted renewed calls for an increase in Atlas deployments.

The controversy over estimates arose, in part, because of Allen Dulles's failure to circulate interim intelligence estimates throughout the administration. The CIA officially released NIE 11-4-59 on February 7[98]—more than two weeks after Gates's initial testimony to Congress. Thus when Gates drafted his testimony he based it on intelligence information that differed in content and emphasis from the information in the finished NIE.[99] Gates complained to Eisenhower that "Mr. Dulles . . . had given figures to the Congress that Defense had never seen prior to that time, and that had not previously been disclosed to the Congress." This situation "embarrassed" Gates and "troubled" Eisenhower.[100]

Democratic congressmen, particularly Senators Lyndon Johnson and Stuart Symington, pounced on this confusion and attacked administration defense policy. Their criticism centered on the adequacy of defense programs and the integrity of intelligence estimates. First, they argued that the administration's budget limitations gave the Soviets an advantage in ballistic missiles and maintained the inadequacy of conventional forces. Statements by Air Force generals on the B-70 and Soviet missiles supported the critics and fueled the uproar. The SAC commander, General Power, publicly claimed on January 19 that the Soviets could soon use missiles to execute a first strike against United States nuclear weapons.[101] Maxwell Taylor also testified before Lyndon Johnson's Preparedness Subcommittee that "the trend of relative military strength is against us." Eisenhower, infuriated by the testimony, considered reprimanding Taylor but decided against it because he did not want to make the general a "martyr."[102]

Second, Democratic critics claimed the Eisenhower administration deliberately misrepresented intelligence estimates to make them consistent with budget limits. Their main evidence was the alteration of the estimating process used in NIE 11-4-59 and inconsistencies in administration testimony. On January 27, Symington stated: "The intelligence books have been juggled so that the budget books can be balanced."[103] The president, still harboring ill feelings toward Symington, rebutted these charges at his press conferences and told reporters that he "deplored" the leaking of classified information in the missile gap debate.[104] Less than one week later, he said charges that he had deliberately misled the public on Soviet missiles were "despicable." On February 19, Symington counterattacked: "Above all, they [the American people] are entitled not to be misled by false statements."[105] These charges angered Eisenhower for obvious and understandable reasons. Goodpaster recorded that "the President expressed his deep concern over what men like Senator Symington or Senator Kennedy might do as President. In his opinion these men lack the judgment required for such responsibilities, even though he recognized that Kennedy is a man of some intelligence.... The President found him [Symington] quite lacking in sense of responsibility."[106]

But once again, this debate over defense policy failed to precipitate any significant changes in the administration's budget submission. Congress added only $781 million to the budget, some of which the administration requested in an April supplemental budget submission. Congressional leaders remained unwilling to pass an alternative defense program over the objections of the president. Public opinion also temporarily silenced Eisenhower's congressional critics. Only 21 percent of the public supported increased defense expenditures; moreover, Eisenhower's approval rating remained stable at about 65 percent. There was so little support for increased defense spending in the Senate that Symington did not even introduce an amendment to add $2.6 billion to the budget. The missile gap debate dissipated after the Soviets shot down a U-2 in May 1960. Democrats criticized Eisenhower's management of the crisis, but they could not charge him with complacency. Still, the defense policy debate had satisfied critics in some respects. For Johnson, Kennedy, and Symington, the missile gap debate promoted their presidential aspirations by depicting them as experts in defense policy, and indicting Richard Nixon—certain to be the Republican Presidential nominee—for the administration's laxity. The debate allowed the armed services to state their objections to the Eisenhower program and gave them an opportunity to cultivate allies for the post-Eisenhower period.[107]

The 1960 Presidential Campaign and the FY 1962 Defense Budget

The missile gap debate was renewed in July 1960 when the Democrats nominated Kennedy and Johnson for their presidential ticket. In accepting the nomination, Kennedy evoked the issue when he said that the choice between his "New Frontier" and the program of the Republicans was a choice "between national greatness and national decline."

Kennedy campaign position papers endorsed the two defense programs that had suffered most from Eisenhower's budget restrictions: the B-70 and conventional forces.[108] Campaign rhetoric stressed the general theme of the decline of the nation's power and international prestige, effectively combining the missile gap issue with international crises such as Quemoy, Matsu, and Cuba.[109] Kennedy's defense proposals included airborne alert, accelerated BMEWS construction, development of the B-70, and increased missile procurement—which would add $2 billion to $4 billion to the budget annually.[110] The Air Force tried to influence Kennedy campaign defense positions through W. Barton Leach. The Harvard Law School professor offered Air Force Chief White "a clear channel" for ideas and even allowed White to critique his memos to the candidate: "If there are any matters on which you think he should be informed . . . by all means let me have them." Leach also asked White for a list to give Kennedy of Republicans in the DOD who should be retained in the event of a Democratic victory.[111] Even without Leach's efforts, Air Force interests were well represented, when Kennedy appointed Symington to chair a report on DOD organization.[112] When the voters elected Kennedy and Johnson, they chose two of the Eisenhower administration's most vocal critics. By rejecting Nixon and the Eisenhower approach to defense, they ensured that Eisenhower's final defense budget would be only an interim document left for the next president.

When the Eisenhower administration began to formulate the FY 1962 defense budget in the summer of 1960, two study groups were evaluating U.S. limited-war capabilities, which had been neglected by the New Look and massive retaliation. In August, a special PSAC panel completed a year-long study on limited war, finding that "our present and planned capabilities fall far short of what could be achieved with present funding and force levels." In its report and in a subsequent meeting with President Eisenhower, the panel carefully avoided advocating increased expenditures, although its recommendations, if implemented, would obviously necessitate increased spending.[113] A CIA-State-Defense interagency panel on U.S. limited war capabilities in various theaters—with a particular emphasis on Laos and Vietnam—found that forces would be "generally adequate" if the United States "promptly" mobilized, lifted expenditure ceilings, and increased airlift capabilities and military production. The interagency panel also found that some limited conflicts "degraded" the nation's capacity for fighting a general war. The report's "conclusion," according to the JCS, "clearly indicates that the United States does not have forces in being adequate to cope with all envisaged limited war situations." The JCS and Defense Secretary Gates believed that the report, because of its narrow assumptions, should not influence policy or budget decisions.[114] But the numerous meetings on the report held by the Planning Board and the NSC in November and December influenced administration decisions on Laos and the defense budget at least indirectly.

Defense Secretary Gates directed the armed services to develop four different budgets for FY 1962, each based on different expenditure levels: no increase from the previous year, a 5 percent increase, a 5 percent decrease, and unlimited funds. The "unlimited" budget allowed the services to submit their "wish lists" to the secretary even though they were unlikely to be filled. The unlimited budget requests by the four services totaled more than $50 billion because of outrageous procurement submissions. The Air Force submitted a five-year plan for the purchase of 142 B-70s costing a total of approximately $5.5 billion. It also recommended a Skybolt cruise missile program requiring $2.1 billion over five years.[115] The task of composing a defense budget from these four service budget submissions fell to the secretary of defense, just as it had in the previous three years. Rising fixed costs in personnel, maintenance, and operations (more than $23 billion for FY 1962) restricted Gates's latitude in "balancing" between the services as well as across different missions.[116]

The PSAC also reviewed the service budget submissions and found four major problems: (1) too many strategic nuclear weapons under development drained resources from other missions; (2) SAC bomber vulnerability was not being reduced quickly enough; (3) limited war forces suffered from "serious deficiencies"; and (4) the armed services were decreasing financial support for both basic and applied research. The PSAC failed to mention that the administration's emphasis on massive retaliation and its fiscal constraints played a major role in creating these problems. The PSAC suggested a series of programmatic changes aimed at reducing them, which included putting one-eight of the B-52 force on airborne alert, reevaluating the Titan and Minuteman force objectives, reducing Polaris vulnerability by increasing missile range and decreasing engine noise, keeping the B-70 development on a prototype basis, canceling Skybolt, and reducing active defenses.[117]

Gates completed his budget and discussed it privately with the president and other senior aides on December 5. He explained that rising fixed costs had led to slower procurement and even undermanned units in every branch of the armed services. "As a result," Gates told Eisenhower, "very little that is really new or that goes beyond present programs is being bought." Eisenhower then quizzed Gates on the B-70. When Gates indicated lukewarm support for the program, the president responded that the aircraft "may be obsolescent as a military weapon before we begin to have it available" because of Soviet air defenses. Other manned strategic programs received even less support from Gates. He authorized no new funding for Skybolt because of technical problems although its development would continue, made possible by stretching out the previous year's appropriation. Gates called the nuclear-powered aircraft "a national disgrace" but would fund it at a very low level. With regard to airborne alert, the Office of the Secretary of Defense (OSD) reduced the Air Force request from $260 million to $70 million, a change that would help stockpile spare parts. The Chiefs approved this reduction, CJCS Lemnitzer told the president, because they "have great reservations about the air alert since it is bound to result in simply wearing out a lot of airplanes." Finally, Gates proposed increasing the Polaris force to 24 submarines (19 fully funded and 5 with long-lead-time funding only) and programming a 540-missile Minuteman force by mid-1964. President Eisenhower approved these proposals but asked Gates to guard against duplication in missiles.

Overall, Gates recommended a $42.9 billion defense budget for FY 1962. Budget Director Stans called it "the toughest, tightest review in the Defense Department." The budget's political viability in Congress, could be increased, Gates believed, if minor increases were made in funding for ballistic missiles. The president rejected this course, employing logic he had used so many times in the past eight years. Goodpaster recorded:

> The President noted that this would, however, inevitably set a "new plateau" for the next year and thereafter. He stressed that our problem is keeping the economy sound for another ten years. He said we have constantly got to ask ourselves whether we are cutting out everything that can be cut out. For example, he is clear in his mind that the only way we are going to win the present struggle is by our deterrent. . . . More and more the matter is a question of big war and the deterrent. If we try to "copper every bet" we will get nowhere. Similarly, simply imposing higher taxes is not an answer since it will have a depressing effect on our economy.[118]

But with the inauguration of a new administration determined to increase defense spending only a month away, Eisenhower's lecture seemed to ring hollow.

On December 8, Gates and Douglas briefed the NSC on the adequacy of defenses and presented a draft defense budget for FY 1962. Although they attested to the general sufficiency of defenses, they also highlighted several long-term problems in both nuclear and conventional forces. Advanced weapons, Douglas said, have "become more costly each year, and so it has become increasingly difficult to accomplish modernization within available resources." With respect to limited warfare, he predicted that "dependent upon the location and size of force required, we would be hard pressed to execute limited military operations simultaneously in two or more areas of the world and maintain an acceptable general war posture."[119] But President Eisenhower challenged the thesis that limited wars would not escalate to a general war—just as he had during the 1959 BNSP debates: "It [is] becoming increasingly dangerous to assume that limited wars [can] occur without triggering general war." Moreover, "our principal effort should be devoted to convincing the U.S.S.R. that no matter what the Soviet Union does, it will receive a rain of destruction if it attacks the U.S. . . . [A]ll other military matters must remain secondary to the overriding importance of deterrence." With some satisfaction, Eisenhower noted that Kennedy had told him that, as a result of a recent SAC briefing, the president-elect now believed that U.S. nuclear capabilities and targets might be excessive.[120]

Gates then presented the defense budget for FY 1962 to the NSC, explaining the constraints that had influenced his proposal. The Army and Navy had complained, Gates said, that the budget provided inadequate funding for procurement, modernization, and personnel. After he finished, Eisenhower allowed each service secretary to present his case for increased expenditures. The president listened to their arguments but challenged them on only a few points. After each service secretary had spoken, Eisenhower confessed that he "hesitated to oppose his dedicated old associates in the military services." He asked them to remember all the economic factors that he had to balance when formulating the final budget. The military must be able to deter the Soviets, but defense expenditures must also be limited. Budget Director Stans agreed and said he supported the budget because it "spread dissatisfaction widely." With the president having remained true to his budgetary convictions, deliberations on his final defense budget came to a close.[121]

As President Eisenhower wrapped up his official duties in January 1961, he took the opportunity to criticize the missile gap thesis, which had bedeviled him for the last three years. In his State of the Union message on January 12, Eisenhower said: "The 'bomber gap' of several years ago was always a fiction and the 'missile gap' shows every sign of being the same." But the departing president saved what would be his most memorable statements for his Farewell Address on January 17—the day after he delivered the budget for FY 1962 to Congress. In a message broadcast on national television and radio, Eisenhower warned that the national security apparatus necessary for fighting the Cold War also threatened the nation's political virtue and liberty: "In the councils of government we must guard against the acquisition of unwarranted influence, whether sought or unsought, by the military-industrial complex. The potential for the disastrous rise of misplaced power exists and will persist." He then described a second "complex" that threatened America: the government-university scientific complex. "The prospect of domination of the nation's scholars by federal employment, project allocations, and the power of money is ever present and is gravely to be regarded."[122]

Scholars and journalists have often cited Eisenhower's military-industrial complex as a warning to future generations. But the debate over of defense budgeting and the existence of a missile gap suggests that the speech was an explanation as well as a warning.[123] In describing the military-industrial complex, Eisenhower was offering the public an explanation of why the political forces organized against his defense program during the missile gap period. After all, was it not the military and arms manufacturers who wanted to break his expenditure ceilings? Criticism by Kennedy, Johnson, and Symington indicated the power these institutions had accrued and the influence they exerted on ambitious elected officials. Symington's association with a Convair vice-president confirmed this. Although Eisenhower had little respect for the politics practiced by these men and often referred to them as "demagogues," he cast his military-industrial complex reference as a warning rather than an explanation out of respect for the institution of the presidency and in accordance with his policy of "not engaging personalities."[124] This speech was President Eisenhower's final performance in the missile gap drama.

The debates over defense budget formulation reveal President Eisenhower's goals as well as the strengths and weaknesses of his leadership style. The attempt to balance security programs and economic performance exacted a heavy toll from the president. He feared that excessive defense spending might lengthen or worsen the recession of 1957–1958. Attacks on Eisenhower's defense budget as well as on his cautious economic policies created a highly charged political environment and made defense budgeting problematic. John Sloan writes: "The constraint he [Eisenhower] exercised over the military budget following Sputnik may have been the most remarkable budget achievement in recent U.S. history. No other politician could have done it."[125]

The stability of defense budgets conceals the important changes which occurred during the missile gap period. The justification for increased defense spending came from several sources: intelligence estimates predicting Soviet missile superiority; increased concern about the threat of conventional warfare in Europe and the developing world; and the simultaneous development of numerous advanced weapons systems. Budget priorities had to be altered for Eisenhower to achieve his goals of economic stability and stable defense spending. The first, and easiest, choice he made was to apply the New Look formula, which emphasized nuclear weapons at the expense of conventional forces. This conformed to the administration's BNSP papers, which continued a national strategy of massive retaliation. Thus, from 1957 to 1961, the strength of conventional forces deteriorated further because of declining personnel levels and weapons procurement that did not even keep pace with attrition. But, extended application of the New Look could not, by itself, produce stable budgets. Consequently, Eisenhower attempted to control defense spending by adjusting strategic weapons programs as well. A few strategic weapons were eliminated (the nuclear-powered bomber), a vast number were reduced to insignificance (the B-70, Skybolt, and Nike-Zeus), and even successful programs were restrained (ICBMs, Polaris, BMEWS). Thus, the content of the defense budget changed dramatically even though the budget level remained relatively constant.

This chapter illustrates the paradoxical nature of Eisenhower's leadership. Within the executive branch, he deftly employed his leadership and management skills to develop budgets that met his goals. Several of his techniques stand out. He concentrated defense budget decision making in the hands of a few civilians—Stans, the secretaries of

Defense and Treasury, and the JCS chairman. This concentration allowed Eisenhower to control the "action-channel" for budget decisions. In this way, he cut out the armed services and even the entire NSC from the important budget decisions that transformed service requests into a budget. By the time the NSC reviewed the budget, Eisenhower and the defense secretary had resolved the major issues. By limiting the participants in the action-channel, Eisenhower effectively limited the bureaucratic politicking within the administration over the budget. Eisenhower also cultivated the advice of the PSAC to avoid overreliance on advice from the services and DOD. Scientific evaluations of weapons systems also enabled the president to keep the secretary of defense "honest" in defense budgeting. The White House forced many weapons cuts on the DOD based on PSAC scientific evaluations. By developing the weapons expertise of non-DOD officials and structuring the decision process, Eisenhower manipulated his administrative machinery to produce defense budgets that remained stable, yet still expanded missile production to keep pace with and ultimately overtake the Soviet Union—truly a remarkable feat of presidential policy making.

In contrast, the record of Eisenhower's leadership outside the administration is mixed. This assessment is based on several measures. Congressional leaders never passed any major revisions in the defense budget over his objections. Public opinion consistently gave him high approval ratings while opposing higher defense spending. Certainly, Eisenhower's personality and his reassuring status quo policies resonated with the American public. But Eisenhower's leadership in the public realm during the missile gap period was deficient in several respects. His unwillingness to revise massive retaliation or to increase defense spending allowed critics to charge him with complacency about the Soviet threat. Eisenhower never publicly articulated a position that would force his critics to drop the missile gap issue once and for all. According to McGeorge Bundy, although Eisenhower "did sensible things and resisted foolish ones, he allowed the ensuing public argument to be led by men who did not understand matters as well as he did."[126]

Four factors inhibited the White House from developing an alternative argument concerning the missile gap issue that would silence Democratic senators. First, since the NIEs predicted a Soviet ICBM advantage, the administration was forced into the precarious position of admitting that a gap would exist but insisting that *it would not be strategically significant.* Second, new intelligence data actually complicated the administration's political problems. Although this intelligence information was spotty and inconclusive, it still allowed the administration to advance the probable onset of a missile gap farther into the future. Democrats responded by charging that the administration either had failed to estimate Soviet missile projections accurately or had deliberately misrepresented the intelligence projections for budgetary reasons. The need for secrecy prevented the administration from informing Congress of the basis for the changes in the NIEs, at least until the U-2 crisis in May 1960. Third, President Eisenhower's leadership was undermined by the administration's political bungling and occasional missteps: administration rhetoric frequently alarmed rather than soothed the public; exclusion of the armed services from budget deliberations precipitated end runs to Congress; administration officals offered contradictory congressional testimony due to miscommunication between departments. These combined to keep the administration off balance politically, so to speak. It seemed as though the Eisenhower administration spent so much time trying to recover from these self-inflicted wounds that it never mounted a clear and consistent defense of its program.

Fourth, and most important, Eisenhower's inability to articulate an alternative argument can be attributed to the leadership style he preferred. According to Greenstein, Eisenhower's leadership style emphasized organization, delegation, effective use of language, personality analysis, an unwillingness to "engage personalities," and a "hidden-hand" persuasion technique.[127] Eisenhower's leadership skills served him well, especially in the process of formulating a defense budget. But in the missile gap period, this leadership style was inappropriate for maintaining the support of opinion elites and the public. Eisenhower called Symington, Johnson, and Kennedy demagogues who attempted to manipulate national security for personal political gain—but he did so in the Oval Office, rarely in public. The political climate required a forceful president who confronted his critics, attacked their policies and motives, and could justify his program in the eyes of the public. In effect, Eisenhower needed to abandon his "hidden-hand" style and take up the political cudgel, even if it meant "out-demagoguing" the Democrats. But with the term "politician" becoming increasingly pejorative, Eisenhower eschewed the public-political leadership role, thus allowing his critics tremendous latitude in the political arena.[128]

The debates over budget formulation in the missile gap period can provide an important lesson for both scholars and presidents: politicians need to exercise political leadership and executive leadership simultaneously if they are to achieve policy goals. But the political milieu may be such that a leadership style appropriate in one arena may be entirely inappropriate in another. Scholars should recognize this whenever they pronounce judgment on presidential effectiveness. Presidents also need the sophistication and acumen that will enable them to adjust their leadership style when necessary.

[5]

Strategic Force Planning during the Missile Gap Period

The Eisenhower administration's uncertainty over present and future Soviet strategic nuclear capabilities in October 1957 was matched by the uncertainty existing about U.S. strategic nuclear forces. The advent of the missile age raised questions concerning the nation's capability to deter and retaliate despite its large nuclear force. With Soviet forces in a state of transition, the administration faced two central issues when planning strategic forces. First, what weapons programs should be undertaken for the short term to ensure deterrent capabilities even if the Soviets achieved some missile advantage? Second, what type of force would provide the best deterrent once each superpower possessed a large missile arsenal? The force planning decisions made by the Eisenhower administration during the missile gap period established the foundation for America's strategic nuclear forces for decades to come.

The missile gap period was a time of major change for U.S. strategic nuclear forces even though defense spending remained stable. In the three years after Sputnik, the American strategic nuclear triad emerged as the Eisenhower administration programmed large ICBM and SLBM forces to add to SAC's bombers. In October 1957, U.S. strategic nuclear forces consisted of 340 long-range bombers (B-36 and B-52), 1,285 medium-range B-47 bombers, while another 130 ICBMs and 48 SLBMs (on 3 Polaris submarines) had been approved. When Eisenhower left office in January 1961, the deployed strategic nuclear forces consisted of 538 long-range B-52 bombers, 1,178 B-47s, 12 Atlas ICBMs, and 16 SLBMs on 1 Polaris submarine. But the administration had also programmed a large missile force: 810 ICBMs (130 Atlas, 140 Titan, and 540

Minuteman) and 24 Polaris submarines that would carry 384 SLBMs. In addition to this triad, the administration shaped the nation's strategic force posture through its decisions about *which weapons not to develop and deploy*.

The Eisenhower administration's role in constructing the strategic nuclear triad has been underappreciated. However, the administration's decisions during the missile gap period established the size and structure of the U.S. strategic nuclear force posture. Recently declassified documents enable scholars to reconstruct the policy process that took place within the administration as well as to gain some understanding of the reasons for the administration's choices.

The first section of this chapter focuses on manned bomber programs, which included airborne alert and dispersal, as well as on the development of air-to-surface missiles and new aircraft. The second section concerns decisions relating to first-generation missiles (Atlas and Titan ICBMs) and second-generation missiles (Minuteman ICBMs and Polaris SLBMs). In making decisions in each area, the Eisenhower administration sought to make the bomber force secure immediately and to work toward developing a militarily useful missile force. Long-range missiles would eventually supplant bombers in strategic missions, although bombers would not be replaced completely. This approach to strategic nuclear force planning alienated both the Air Force and the missile gap critics. The Eisenhower administration's reluctance to support any manned bomber programs other than those necessary to ensure deterrence for the short term threatened the central component of the Air Force's organizational essence. Moreover, the administration's decision to delay a large missile expansion until more militarily reliable second-generation missiles became available provided fodder for the missile gap critics. Opponents of the Eisenhower approach to strategic nuclear force planning easily found common ground, especially when the budget constrictions are taken into account.

Manned Bombers in the Missile Age

Manned bomber aircraft had been the sole mechanism for delivering U.S. strategic nuclear forces since World War II and formed the basis for Air Force doctrine and organizational behavior. But Sputnik and the coming of the missile age raised two questions about the utility of manned bombers in the immediate future and in the long term. First, Soviet deployment of a large ICBM force posed a potential first-strike threat which might undermine America's capability to retaliate—and in turn, to deter. In October 1957, SAC's 1,625 bombers were concentrated at only 29 bases (although SAC planned to disperse the force across 46 bases in 1958).[1] These bombers seemed vulnerable because they constituted a small number of targets for the Soviets and because none of the facilities had been designed to withstand even a small nuclear detonation—that is, they were "soft" targets. These bombers would become vulnerable to a first strike in several years if the Soviets instituted a crash ICBM program (see Chapter 2).

Second, vulnerable manned bombers might ultimately be replaced by ICBMs or SLBMs that offered short flight times, quick response, and relative invulnerability both before and after launch. This possibility had worried top Air Force leaders for years. Even before Sputnik, they had tried to protect manned bombers by restricting ICBM development.[2] But the public uproar over Sputnik threatened to create an unstoppable momentum in favor of a shift from bombers to missiles. Air Force officials feared their organization would be transformed from one responsible for manned strategic bombing to an organization of passive automatons that fired missiles.

The Eisenhower administration made a series of important decisions relating to manned bombers during the missile gap period. The first group addressed bomber vulnerability in the near term and included programs for dispersal, a Positive Control or "Fail-Safe" program, and ground and airborne alert. The second group concentrated on expanding bomber capabilities by arming B-52s with nuclear ASMs and by developing new bomber aircraft (the B-70 and the nuclear-powered bomber). The Air Force, eager to protect its organizational essence and its dominance over strategic missions, enthusiastically advocated both types of measures. DOD civilians and the other services generally accepted programs to reduce bomber vulnerability in the near term. But there was less unanimity on the issue of expanding bomber capabilities. President Eisenhower, most of his senior officials, and the other services opposed these programs largely on the grounds that bombers were a less effective and less economical way of delivering nuclear weapons than missiles. Such open questioning of the Air Force's organizational essence contributed to the intense interservice rivalry over force planning and affected policy debates.

Protecting Manned Bombers in the Near Term

Ensuring SAC's retaliatory capability in the near term necessitated changes in bomber deployment and operations to preclude a successful Soviet attack: increasing the number of targets the Soviets would need to strike, dispersing bombers, and placing bombers on alert so they could become airborne before the Soviet attack arrived.

The Gaither panel considered SAC vulnerability to be the most pressing problem facing the nation. In its report it recommended spending an additional $8.3 billion over the next five years for SAC alert, dispersal, base hardening, warning systems, and active defense of bases.[3] Two of the panel's most influential members, William Foster and Robert Sprague, were particularly disturbed by an SAC exercise in September in which six hours elapsed before any aircraft became airborne. In a private meeting with President Eisenhower on November 7, the two explained that SAC might be unable to retaliate against a Soviet surprise attack in periods of low international tension if it did not receive sufficient warning: "We cannot assume we could lay down a substantial retaliatory attack."[4]

Unbeknown to the Gaither panel, the administration had already begun discussing the reduction of SAC vulnerability.[5] A month earlier, CINCSAC General Thomas Power had initiated a ground alert or runway alert program in which 134 bombers would be able to take off within a period ranging from 30 minutes to 2 hours after receiving warning of an attack.[6] Spurred on by the Gaither report, SAC expanded the ground alert force to 157 bombers in January 1958 and planned to increase it to 515 bombers by mid-1959.[7] Having faced little opposition, SAC achieved these objectives; the ground alert force grew from 11 percent of bombers in 1957 to 20 percent in 1959 and 33 percent in 1960.[8]

Intelligence evaluations of the nation's warning system and the Soviet threat prompted Power in late 1957 to institute more extreme measures to protect SAC bombers. The Air Defense Command had informed Power that a Soviet bomber attack at high and low altitudes (more than 50,000 feet and less than 2,000 feet) might elude much of the warning system. The warning then available for SAC bombers was only: 1–2 hours for 11 bases; 30 minutes to 1 hour for 10 bases; and less than 30 minutes for 11 bases.[9] Further, the CIA claimed that the Soviets had retrofitted some submarines with cruise missiles that had a range of 500 miles. Half of SAC's bases in the United States would receive no warning from this relatively unsophisticated system.[10] Recognizing that warning problems would become much worse once the Soviets began deploying strategic missiles, Power ordered that the SAC ground alert force reduce its reaction time to less than 30 minutes.[11] He also developed plans to put one-third of the SAC force on 15-minute ground alert in 1960—when it was anticipated that the Soviets would possess a large ICBM force.[12]

Keeping bombers in a state of high alert increased the risks of an accidental or unauthorized nuclear strike, especially in a period of inadequate warning systems and before Permissive Action Links. As a result, SAC and the Air Force developed a "Positive Control" or "Fail-Safe" program, which enabled ground alert bombers to be launched in a crisis or after warning of an attack. The aircraft would fly to predesignated points along the "Positive Control Line" (PCL) that stretched across the Arctic Circle, 300 miles from Sino-Soviet borders to avoid radar detection. Bombers would wait at the PCL for an order to proceed to targets and would return to U.S. bases if one was not received. Because bombers might be airborne for hours before receiving the "Go" code, each aircraft carried plans for striking three different targets, depending on the amount of fuel that remained.[13] Testing showed that 95 percent of the bombers would receive the code if it were sent. On March 1, 1958, CSUSAF White ordered SAC to conduct Positive Control operations in case of a "Defense Emergency."[14] When Air Force officials briefed the NSC on April 3, 1958, President Eisenhower and others wondered whether launching bombers under the Positive Control program in a crisis might precipitate a nuclear war because of Soviet misperceptions of U.S. intentions.[15] The Army and the Navy shared these concerns but the JCS eventually decided that this risk of accidental war was small and was outweighed by the program's benefits. DCI Dulles said that the anticipated Soviet reaction did not require any special analysis by the intelligence services.[16] By mid-1958, the Eisenhower administration had begun to take measures to reduce bomber vulnerability by implementing dispersal, ground alert, and a Positive Control launch option.[17]

As implementation of such measures progressed, SAC developed a program of airborne alert in which a small number of bombers armed with nuclear weapons would be in the air at all times, ready to strike the Soviet Union if they received the "Go" code. After keeping one B-36 on airborne alert in the spring,[18] SAC completed a plan for a limited two-part test and presented it to Eisenhower in July. In the test's first phase, which would begin on September 15 and last several months, SAC would maintain three B-52s armed with nuclear weapons in the air at all times. The planes would fly missions lasting twenty hours and could strike the Soviet Union at any time except during the last two hours. The second phase of the test program would begin on March 1, 1959, with operations contingent on evaluations of the first phase. Cumulatively, the plan called for almost 750 B-52 airborne alert sorties over a period of seven months. SAC would be flying "war-ready" nuclear weapons for the first time; the risk of premature nuclear detonation and other accidents had prevented the transportation of assembled nuclear weapons in the past. The Air Force claimed that advances in sealed pit warhead technology now reduced the likelihood of an accidental nuclear explosion to nearly "zero."[19] With Eisenhower's approval, SAC carried out the first phase of the test in the fall of 1958 under the code name "Head Start I."

Even though the administration had approved the airborne alert exercise in August, many top officials favored other options for reducing SAC bomber vulnerability, such as building a second runway at SAC bases. General Twining believed that "the survivability of the force would be almost doubled by providing dual runways," which could reduce the time needed to launch the ground alert force. Eisenhower expressed enthusiasm for the concept; he told Twining that "we should make all possible arrangements to 'bomb up' rapidly and get the planes into the air." Twining advocated further study because of the proposal's cost ($11.6 million per B-52 base and $7 million per B-47 base).[20] In October, the DOD suggested deferring the proposal until after implementation of the dispersal and ground alert programs because of the costs as well as the fact that the bases would still be vulnerable to a single Soviet nuclear weapon.[21] The DOD nevertheless commissioned a WSEG report on dual runways, probably because of the president's interest in the option.

Only days after the second part of the Head Start exercise began in March 1959, CINCSAC Power requested that the JCS make airborne alert a permanent program. In justifying the request he cited the estimates of increases in Soviet ICBM forces, weaknesses in early warning capabilities, and the Berlin crisis.[22] The request came in the midst of the first missile gap debate, and Power hoped to capitalize on the public controversy.[23] The JCS rejected the request on the grounds that "the tests to date have not provided sufficient data to evaluate the general benefits to be derived from a continuous airborne alert on the remaining alert forces."[24] Several other factors, primarily the high cost of airborne alert and CSUSAF White's interest in augmented SAC dispersal to civilian airfields, weighed heavily in the JCS decision.[25] Airborne alert remained confined to the five B-52s that were aloft from March through June 1959 under Head Start II.

The administration grappled with the idea of building a second or dual runway at bases to reduce SAC vulnerability in the summer of 1959, after the WSEG had completed WSEG-SS-77. To ascertain the strategic value of dual runways, the WSEG estimated the number of aircraft that would retaliate after different Soviet surprise attacks. As discussed in Chapter 2, the WSEG predicted in this report that SAC would be frighteningly vulnerable to Soviet surprise attacks in 1960 and 1961 if the NIEs of Soviet missile strength proved correct. Despite this risk, the WSEG withheld unconditional support for dual runways. Such runways offered relief only for medium-range bombers and tankers that were crowded on a small number of bases. The WSEG recommended building emergency runways at six bases to increase the launch rate of medium-range aircraft by 33 percent during an ICBM attack. Chairman of the JCS Twining informed Eisenhower of the WSEG report in August.[26] There is no evidence that the administration pursued the issue any further, probably due to the growing awareness of B-47 weaknesses and obsolescence (see next section).

At the same time, SAC again requested a permanent airborne alert program, now that Head Start tests had been completed. Growing estimates of Soviet missile forces, according to SAC, suggested the need for an initial program of 6 sorties per day with a gradual increase to 11.2 sorties per day at a cost of between $500 million and $1 billion annually. Air Force Secretary James Douglas endorsed the request but conceded that it could be funded at lower levels.[27] A Joint Staff evaluation admitted that the projected increase in Soviet strategic forces required greater protection of SAC bombers.

> The United States shortly will be confronted with a serious Soviet ICBM threat; by mid-1961 and continuing until perhaps 1964, pre-planned action must be taken to the maximum extent feasible to neutralize this threat to SAC bomber forces. . . . Under conditions of a surprise ICBM attack, the survivability of an effective portion of this long range nuclear retaliatory capability, even under a fifteen minute ground alert posture and the degrees of dispersal and hardening which are economically and operationally feasible, becomes highly problematical by mid-1961.

However, it reported that airborne alert, would be strategically feasible only if a very large force went on alert: "One of the factual conclusions of the Headstart tests was that for airborne alert to be effective, the entire B-52 force must be positioned on a continuous airborne alert posture as opposed to a combination of ground alert and air alert concurrently and/or sporadically." It advised that an airborne alert program for one-fourth of the B-52 force should be developed but should remain "on the shelf" until the president decided to implement it.[28]

Chief of Naval Operations Burke, embroiled in a long struggle with the Air Force over the Polaris (see Chapter 3 and below), took this opportunity to attack the Air Force's continued reliance on vulnerable bombers. He questioned the need to undertake expensive measures such as airborne alert at a time when missile programs could perform similar missions. Burke recommended increased emphasis on "those weapon systems possessing a high invulnerability to surprise attack," that is, Polaris.[29] The costs of airborne alert, he argued, should be recouped by "reducing" B-47 operations. The JCS forwarded a split paper to Defense Secretary McElroy after the newly appointed Army chief of staff, Lyman Lemnitzer, had agreed to a revised version of Burke's memorandum.[30] Eventually the secretary of defense accepted the proposal for an "off the shelf" airborne alert, mostly because of the program's exorbitant cost. This allowed SAC to purchase spare parts that could be stockpiled and to continue a small number of airborne alert training flights. In January 1960, Eisenhower approved continuing the "off the shelf" airborne alert for one-eighth of the B-52 force and indicated a willingness to increase funding if circumstances so dictated.[31]

In the spring of 1960, the administration began to consider additional measures to improve the nation's defense posture after Advanced Research Projects Agency (ARPA) Director Herbert York projected that a possible missile gap might exist by 1961 (see Chapter 2). Air Force Chief of Staff White pressed York to recommend a permanent airborne alert for SAC: "A one-quarter airborne alert with optimum weapon loading could, after degradation for operational factors, put over 300 weapons, or more than 700 megatons of additional yield, on enemy targets. It appears that of all the interim possibilities open to us, no single other action or combination of actions would minimize the threat which you outlined to the extent attainable through airborne alert."[32] While the administration showed little inclination to change its position, SAC's case was bolstered by deteriorating relations with the Soviets after U-2 and RB-47 aircraft were shot down as well as the appropriation by Congress of $85 million more than requested by the administration for airborne alert. With airborne alert training flights due to be reduced from 12 to 6 flights per day, SAC Commander Power hoped that keeping 30 B-52s (one-sixteenth of the force) airborne would "establish airborne alert as a way of life."[33] Defense Secretary Gates became interested in increasing airborne alert flights in July after a discussion with the JCS about improving the nation's military posture. In August, the JCS approved Power's recommendation to increase the number of training sorties to 29 per day.[34] The Air Force continued to plan for having 25 percent of its bombers on airborne alert by January 1, 1962.[35]

By the end of the Eisenhower administration, SAC had flown more than 6,000 airborne alert sorties, even though the program remained at a training status.[36] Bombers were in the air, but only a small number were flying at any given time. There is little evidence the administration discussed the political signals of airborne alert, even though the Soviets periodically complained that U.S. bombers were flying in the Arctic region.[37] WSEG-52, completed in early 1961, stated that "provocations" could arise in a crisis because the Positive Control Line was closer to the Soviet Union than the airborne alert routes. If the Soviets knew the location of airborne alert operations, then they might interpret a decision to send bombers to the Positive Control Line as the beginning of a U.S. attack.[38] The possibility or intensity of "provocations" might be increased by the fact that SAC airborne alert plans did not allow for a partial release of bombers or for recall before they reached the PCL.[39]

The Eisenhower administration and the JCS were reluctant to implement a full airborne alert because of the wearing effect it would have on bomber aircraft that already had suffered structural engineering problems. In late 1957, SAC revised bombing tactics from high- to low-altitude missions in response to Soviet development of high-altitude air defenses. When B-47 crews began training for the new missions in early 1958, SAC suffered 14 accidents and 34 casualties.[40] The Air Force determined that "structural fatigue" had caused the accidents. The B-47, which had been designed for high-altitude missions, simply could not withstand the enormous pressures of high-speed, low-level bombing. The strain of these missions produced cracks in almost one-third of SAC's 1,367 B-47 bombers. The Air Force eventually corrected the cracks and reinforced the wings of all B-47s at a cost of $60 million.[41]

These sudden problems raised doubts about the B-52, since it too was designed for high-altitude bombing. Tests initiated in the summer of 1958 revealed the distressing information that extreme weights carried at low altitudes weakened the B-52's secondary structure. In February 1959, the Air Force and SAC established limits on B-52 use because of various structural problems, including cracks in the skin and ribs on wings, cracked fittings, and "questionable forgings."[42] By the summer, Boeing flight tests of a B-52D had confirmed the Air Force's worst fears. On June 27, a test aircraft crashed while replicating an attack mission because its structure failed to withstand the immense pressures of the flight. When three more such crashes occurred, the Air Force canceled all low-level, high-speed B-52 test flights. The Air Force and SAC began immediately to seek remedies for some of these defects, but the complete results of the structural tests would not be available until 1961.[43]

The problem of bomber structural fatigue had two important effects. First, it precluded airborne alert unless the administration was willing to procure more bombers and spare parts to compensate for the accelerated wear and tear. Second, there was a decrease in SAC's bomber force size. B-47 structural fatigue, coupled with the heavy dependence on tankers and overseas bases, led the administration to retire 179 B-47s in its last two years—even after Eisenhower slowed the retirements in mid-1960 as a signal to the Soviets. The administration offset this decline to some extent by deploying 158 B-52s during the same period.[44]

The Eisenhower administration was receptive to most measures aimed at reducing bomber vulnerability during the missile gap period. The major disagreement over short-term measures arose between the Air Force and the administration over airborne alert. The administration kept airborne alert on a limited basis because of its high costs and the wear suffered by the aircraft but was prepared to expand the program if the international situation warranted. Reconfiguration of SAC operations by initiating dispersal, airborne alert, and Positive Control reassured the Eisenhower administration of SAC's capacity to deter and retaliate in a period of strategic uncertainty.

Expanding Manned Bomber Capabilities

Strategic missile development and deployment by the United States and the Soviet Union threatened the viability of bombers over the long term. Soviet missiles would increase bomber vulnerability; U.S. missiles offered the Eisenhower administration a less vulnerable and perhaps less costly alternative to bombers. The Air Force faced a strong challenge to its organizational essence of manned strategic bombing. Realizing that steps should be taken to preserve this core mission before the United States deployed militarily useful missiles, Air Force officials proposed programs that expanded bomber capabilities. Such changes, they believed, would allow bombers to be competitive with missiles in key criteria such as performance, flexibility, and cost. The proposed programs called for arming bombers with ASMs and developing new aircraft, namely, the B-70 and the nuclear-powered bomber. But President Eisenhower, concerned about the budgetary effects of developing multiple strategic systems, proved to be a skeptical audience.

Air-launched missiles. One Air Force strategy intended to expand bomber capabilities was to develop nuclear-armed cruise missiles and ballistic ASMs. These ASMs offered bombers a "standoff" attack capability, which allowed targets hundreds of miles from the bombers to be struck. Since the targets were likely to be active defenses, the bombers could then proceed on their penetrating mission.[45] Concern about bomber penetration capabilities against Soviet active defenses before the launch of Sputnik led the Air Force, under pressure from CINCSAC LeMay, to issue a contract to North American Aviation for development of the AGM-28 or "Hound Dog" ASM.[46] In the spring of 1958, the Eisenhower administration approved the Air Force's request for $100 million to accelerate the Hound Dog program as a hedge against the projected increase in Soviet missile strength.[47] Deployed in early 1960, the Hound Dog carried a 4 MT warhead, had a range of up to 675 miles, a speed of Mach 2, and a CEP of 1 mile. The missile had some reliability problems, and it decreased B-52 performance since each pair of Hound Dogs weighed 20,000 pounds. Still, the missile encountered little resistance from DOD civilians or the other services because it increased bomber capabilities with minimal cost and effort and made the B-52 more flexible, especially when used in alerts.[48] The Hound Dog was soon deployed in large numbers: 54 in 1960; 230 in 1961; 547 in 1962;

and 593 in 1963. In fact, 308 Hound Dogs were still in service as late as 1976.[49]

In early 1959, just as North American Aviation prepared for the first Hound Dog flight test, the Air Force proposed a follow-on missile, the advanced air-to-surface missile (AASM) for U.S. and British bombers. This missile, according to Air Force plans, would carry a 0.5 to 1.0 MT nuclear warhead a distance of between 1,000 NM and 1,500 NM at speeds of up to Mach 5 and would have a 3,000 foot CEP. These performance criteria exceeded those necessary for destroying Soviet active defenses in penetrating missions. In effect, the Air Force was attempting to increase B-52 standoff capabilities to such an extent that manned bombers would be as cost-effective as ICBMs and SLBMs. The DOD director of guided missiles, William Holaday, recognized this attempt and sought a JCS evaluation before he would approve the missile's development.[50]

The AASM fell victim to the interservice rivalry that divided the JCS during the missile gap period. The Army and Navy criticized the Air Force's cost estimates and viewed its assessment of the AASM's operational characteristics as overly optimistic. More important, they argued that the AASM would be of little value after 1962 because of SLBM and ICBM deployments: "Development of this ASM would extend the period of dependence on the manned bomber at a time when major emphasis in the U.S. retaliatory forces should be placed on ballistic missiles. Not only would R & D funds for the AASM have to be diverted from advanced ballistic missile programs, but also, once developed and produced, there would be a natural reluctance to abandon the large investment in this weapon by converting to missiles."[51] CSUSAF White asserted that the AASM would aid bomber "reaction time and penetration capability."[52] On April 17, 1959, CJCS Twining forwarded the split to the OSD and recommended approving a program for research and development only, which Defense Secretary McElroy subsequently did.[53]

The AASM, now named the GAM-87A Skybolt, encountered criticism from many quarters in the fall of 1959. As Hound Dog entered production, the Air Force argued during the FY 1961 budget negotiations in favor of building as many Hound Dogs as possible even if that meant slowing B-52 procurement.[54] At the same time, the results of a DOD study of Skybolt conducted by a panel chaired by James Fletcher indicated that Skybolt's operational requirements would require a more demanding research and development effort than the Air Force had anticipated. The Fletcher panel believed that as a result, Skybolt would cost between $1 billion and 2 billion—three times more than the Air Force had estimated.[55] This negative evaluation of Skybolt, coming as it did at the beginning of Hound Dog production, led some to argue for development of a modified Hound Dog instead of Skybolt. The Air Force now occupied the difficult position of having to defend two analogous weapons systems strongly enough to protect each but not so strongly that one would undermine the other.[56]

With the future of Skybolt in doubt because of rising development costs and shrinking support, the Air Force decided to reduce the missile's operational goals in late 1959, after the WSEG had concluded that AASMs could still be an economic alternative to ICBMs in the 1960s.[57] An Air Force study group recommended deploying Skybolts with ranges of either 600 miles or 1,000 miles (carrying 1.0 MT or 0.4 MT warheads respectively) and a CEP of 1.5 miles. The study group projected production of 1,000 missiles with total program costs of $893.6 million. The Air Force hoped this revised program would encourage the OSD to release $35 million that had been impounded from FY 1960 appropriations.[58]

Many civilians in the OSD and the White House remained unconvinced of Skybolt's utility. The PSAC Missiles Panel began to focus on Skybolt, although the panel's technical adviser, George Rathjens, felt that "it may be that the pressures to continue down the present path will be overwhelming, even if we should recommend against it."[59] In its May 1960 report to Kistiakowsky, the panel concluded that "we are not yet persuaded that the Sky Bolt [*sic*] has great merit."[60] The results of a full Missiles Panel review several months later confirmed these doubts: Minuteman and other ballistic missiles would be able to perform missions identical to those of Skybolt more reliably and at a lower cost. Further, Skybolt might not provide the United States with any capability that could not be attained by upgrading Hound Dog. The Missiles Panel advised Kistiakowsky, with some concurrence from the OSD, to recommend to Eisenhower that he terminate Skybolt immediately. The PSAC pressed for Skybolt's cancellation in December 1960 as the administration was completing its defense budget for FY 1962.[61] Defense Secretary Gates decided not to request any new funds for Skybolt but avoided canceling the weapon entirely by reprogramming $70 million from the previous year's appropriation to help cover the projected Skybolt development costs of $149 million in FY 1961.[62]

The indecisive outcome with regard to Skybolt resulted, in part, from

British pressure on the Eisenhower administration. Whereas Skybolt benefited the United States only marginally, the British counted on it to preserve an independent nuclear deterrent into the future and had even canceled its own "Blue Streak" missile because of it. Deputy Defense Secretary Douglas confided to Kistiakowsky that "he was also opposed to the Skybolt, but that the support from the Air Force was very strong, and the British were putting pressure on because for political reasons they wanted to have a 'ballistic missile' in view of the fiasco of the Blue Streak."[63] Several times in 1960, the Eisenhower administration assured the British of continued Skybolt development—contingent on the overcoming of technical hurdles. As a quid pro quo, the British granted the United States additional submarine-basing rights for Skybolt development.[64] Strengthening the British deterrent by the transfer of Skybolt missiles—less the nuclear warheads, of course—reinforced Eisenhower's general policy goal of promoting "nuclear sharing" among the European Allies. But Skybolt was hardly a sturdy vehicle for promotion of such a sensitive policy because of its technical problems. As the PSAC Missiles Panel noted in its July 1960 report, "the Panel is aware of the fact that cancellation of Skybolt may possibly result in embarrassment to the United Kingdom, in view of the fact that its development appears to have been used as a rationale for canceling Blue Streak. . . . It may be noted . . . that various conversations, agreements, and the interchange of personnel between the U.K. and the U.S. are having, and will continue to have, the effect of solidifying and deepening the U.S. commitment to the U.K. in connection with this program as time progresses."[65] Consequently, the Skybolt program lingered on at the end of the Eisenhower administration. Its technical problems made an upgraded Hound Dog missile an attractive and possibly cost-effective alternative. Skybolt had few supporters other than the Air Force and British.

The B-70 bomber and the nuclear-powered bomber. The most significant Air Force efforts to maintain manned bombers into the missile age rested on the development of new aircraft that could be distinguished from missiles in terms of either capabilities or cost-effectiveness (or both). The Air Force hoped to preserve its organizational essence of manned strategic bombing with the development of revolutionary aircraft such as the B-70 bomber and the nuclear-powered bomber. Both programs were begun several years before Sputnik. For the Air Force, the Sputnik launch marked the beginning of a race against time. The strategic and economic viability of new bombers would have to be proved before accelerated missile programs reached the deployment stage. If missiles completed testing rapidly and entered production sooner than anticipated, then DOD civilians and top officials in the other services would have a ready excuse for cutting or even canceling the manned bomber development programs. Under such circumstances, manned bombers might be relegated to a supplemental or even marginal strategic role in the future. But the B-70 and the nuclear-powered bomber were hardly well situated in October 1957 to prevent this.

Air Force development of a follow-on to the B-52 had begun almost three years before Sputnik. To increase the chances of funding, the new bomber would have to have performance capabilities that distinguished it from existing bombers (B-47, B-52, and B-58) as well as from first-generation missiles. Based on research by the National Advisory Committee for Aeronautics, the Air Force decided that the bomber should be propelled by a special new chemical fuel, be able to fly at altitudes of more than 70,000 feet to avoid defenses, and be capable of very high speeds for cruising and target runs (Machs 0.9 and 3.0, respectively).[66] The Air Force believed that the high speeds and high altitudes could compensate for the problems that existing bombers had in conducting penetrating missions, yet retain the payload and accuracy advantages of bombers over missiles. The new bomber, the B-70, also had many problems—all related to the Air Force's technical specifications and exacerbated by its management plans. In its first two years, numerous readjustments were made in the B-70 development program, for scientific and fiscal reasons.

The initial post-Sputnik defense expansion brought new activity in, and hope for, the B-70. On December 23, 1957, North American Aviation received the prime contract for development of the B-70, and General Electric received the contract to develop the chemical high-energy fuel engines. The Air Force accelerated development timetables to achieve flight by the end of 1961, with initial deployment by August 1964. It estimated that these actions would increase program costs by $165 million and that the total cost of the entire 45-aircraft program would be $2.3 billion.[67] But the administration's hesitancy about B-70 funding during the October 1958 FY 1960 budget deliberations prompted Air Force Chief White to delay the projected dates for initial flight and deployment. The B-70 received only $221 million in the administration's budget proposal for FY 1960.[68]

The B-70 development program encountered a number of serious problems in 1959. Failures in developing high-energy fuel engines forced the Air Force to switch to conventional fuel engines.[69] The Eisenhower administration's tight budget ceilings for FY 1961 brought more bad news. As the Air Force formulated its budget, it realized that it could not afford to develop both the B-70 and a new fighter, the F-108. The Air Force decided to cancel the F-108, but this drove up B-70 costs, since the development costs for a number of common subsystems had been divided between the two aircraft.[70] Again, the Air Force was forced to reschedule initial B-70 deployments farther into the future.

With the B-70 program mired in problems, the Air Force tried to formulate a strategic rationale which could justify the aircraft development in the upcoming FY 1961 budget negotiations with the administration. The Air Force case rested on four arguments. First, a strategic triad that included B-70 bombers presented the Soviet Union with additional defensive considerations. Second, only the B-70 could perform special tasks such as strategic reconnaissance and destruction of hardened targets. Third, continuing B-70 development reduced the risks of failures in missile development. Fourth, the human element in bombers represented an important operational advantage over missiles. As Power told White: "Another major requirement for manned penetrators lies in the fact that you cannot put eyeballs on a missile warhead. Only manned penetrators can bomb poorly located and ill-defined targets."[71]

The B-70 dominated the Eisenhower administration's final FY 1961 budget deliberations at Augusta, Georgia, in November 1959. In the first of a series of meetings, Eisenhower expressed to key advisers his doubts about the Air Force's need for a manned bomber–reconnaissance capability in the missile age: "If we place ourselves in 1965, then in those 6 years we should know whether missiles are as effective as we now believe. If they are effective, there will be no need for these bombers. . . . The Air Force must make up their minds." Defense Secretary McElroy tried to justify the B-70 on the basis of its civilian applications but the president "sharply" objected and claimed to be "allergic" to it. The B-70 received tentative support from Gates and Twining, while science adviser Kistiakowsky questioned the plane's ability to evade Soviet active defenses. Twining offered the B-70's capacity to destroy Soviet mobile ICBMs as another justification for the program. According to Goodpaster, this prompted a stinging rebuke from Eisenhower:

"The President said that, if they think this, he thinks they are crazy! . . . To spend $385 million on a vehicle which would never be useful militarily is foolish in his opinion. We are not going to be searching out mobile bases for ICBMs, we are going to be hitting the big industrial and control complexes."[72]

The Air Force presented its case to Eisenhower in a budget meeting with the JCS two days later. White argued for the B-70 on five grounds: other Air Force programs had been cut in order to keep it in the budget; it was "too far" along to cancel without wasting billions of dollars; the United States should not rely exclusively on untested missiles for deterrence; bombers had unique performance capabilities; and bombers had a "powerful psychological impact." But Eisenhower replied that the B-70 "left him cold" militarily and rejected White's arguments. The Air Force chief "begged" to keep the B-70 alive with a "bare minimum" program costing $200 million. Unimpressed, the president replied that missiles would be able to perform identical missions by the time the B-70 was deployed: "We [are] greatly overinsuring our ability to hit an enemy. There is no uncertainty that we would be able to hit his cities. . . . [I find] the missile a cheaper, more effective way of doing the same thing. . . . In ten years the missile capacity of both countries will be such as to be able to destroy each other many times over. . . . We are going overboard in different ways to do the same thing." But White still claimed the B-70 should be funded because it was the only bomber then under development. The other Chiefs with the exception of CNO Burke supported this proposal, although Lemnitzer argued for less than $200 million. The president said he would consider White's proposal, but he compared bombers in the missile age to "bows and arrows at the time of gunpowder." He insisted that "each Chief must look for every possible saving, even driblets. . . . The question is simply one of success in rocketry. This success has made possible and necessary reductions in aircraft programs. It is a change in our thinking."[73]

President Eisenhower revealed his budget decisions to his senior advisers at the final Augusta meeting on November 21, 1959. The president justified limiting B-70 funding to $75 million by arguing that missiles would provide an equally effective yet less costly means of deterrence. He told them that "all we really have that is meaningful is a deterrent. If the Soviets think the B-70 is more effective than missiles, then it has value. If they do not, it is valueless."[74]

Eisenhower's logic for the budget decision realized the Air Force's worst fear—that the manned strategic bomber would be replaced by ballistic missiles. The decision also represented a major attack on its organizational essence. Almost immediately, the Air Force canceled B-70 subsystems development, intending instead to build a single prototype aircraft.[75] General White, the Air Force chief of staff, requested a briefing concerning "How risky is it, from an operational standpoint, to rely on ICBMs as a primary weapon in our deterrent force when we have yet to test the vehicle married with a warhead?"[76]

The Air Force decided that the high stakes demanded an appeal to Congress for a higher B-70 appropriation.[77] As Congress reviewed the budget in January 1960, President Eisenhower tempered his opinion of the B-70 program and came "to the conclusion that continuation of research and development is wise." He expected that Congress would increase B-70 funding and told Twining he would not object so long as the increase did not exceed $100 million.[78] However, others in the administration continued to voice their opposition to the aircraft. Kistiakowsky alerted the president to various problems and claimed that "it is not clear what the B-70 can do that ballistic missiles can't—and cheaper and sooner at that."[79]

The summer and fall of 1960 brought many twists and turns for the B-70 program. The Air Force and North American Aviation signed a development contract for a single experimental prototype (the XB-70) on June 27.[80] Three days later, Congress completed the budget for FY 1961, which included expenditures of $265 million for the B-70—$190 million above the administration's recommendation. Based on the availability of additional funds, the Air Force approved returning the B-70 to "weapon system status." The DOD indicated its support for this action by expanding the program to thirteen test aircraft on August 24. However, the administration impounded $155 million of the appropriation. Kennedy's attacks concerning the missile gap during the 1960 presidential campaign focused attention on Eisenhower's vacillation with regard to the B-70. Kennedy told a campaign audience: "I wholeheartedly endorse the B-70 manned aircraft."[81] The Eisenhower administration released the impounded B-70 funds one week prior to the election, perhaps to boost Nixon's campaign in California, where much of the plane would be built.[82]

At the same time, the Air Force prepared a B-70 proposal for the FY 1962 budget that called for an expenditure of $5.5 billion over the next five years for the purchase of 142 aircraft. The Air Force submitted four budget levels for FY 1962 ranging from $358 million to $586 million, each of which would result in a different operational date for the aircraft. But PSAC staffer George Rathjens advised Kistiakowsky that $490 million was the "most sensible" amount. Aware of Kennedy's campaign promises and the administration's release of the impounded funds, Rathjens was now resigned to production of the B-70: "My belief is that the question of going to production on the B-70 is now largely an academic one and that it cannot be stopped."[83] The PSAC ultimately recommended constructing two to four XB-70s for civilian use since it did not "believe that the B-70 is likely ever to be very useful as a weapon system."[84] Despite his willingness to release the impounded B-70 funds to aid Nixon's campaign, Eisenhower remained skeptical about the aircraft's utility because of Soviet air defenses and U.S. missiles. He told his advisers: "The B-70 . . . is four to five years away, and . . . it may be obsolescent as a military weapon before we begin to have it available."[85] Defense Secretary Gates considered canceling the bomber outright to save $400 million. Ironically, the B-70 was not even raised at the final NSC meeting on the defense budget. The administration's budget requested $358 million, which would be part of a $2.7 billion program to construct twelve XB-70s.[86] This may have been done in anticipation of the new administration's defense priorities.[87]

The Air Force investigated a number of other exotic projects for manned aircraft during the missile gap period, the most important being the nuclear-powered bomber. Nuclear propulsion appealed to the Air Force in the 1940s and 1950s for a variety of reasons, including the bomber's unlimited range and very long times aloft. Such aircraft would not have to depend on overseas bases or air refueling and would be ideal for airborne alert missions and airborne command posts. But the technical obstacles were enormous. The development of a nuclear reactor small enough to fit on an aircraft yet powerful enough to propel it was a major undertaking. Airframes would be needed that could shield the crew from radiation. Even the resolution of these problems would not necessarily result in a militarily useful aircraft. The plane's performance and cost characteristics would still have to be competitive with those of existing conventional aircraft and missiles. However, reactor and shielding weight and costs made this difficult. As a result, the program's status changed many times in the pre-Sputnik years. The program concentrated mostly on building a suitable reactor, although twice it focused on developing a full weapons system.[88]

When the Soviets launched Sputnik I, the Air Force's nuclear-powered bomber program was limited to the building of a reactor unit (in conjunction with the AEC), with little or no commitment to develop a complete weapons system. The Air Force planned to deploy a few nuclear-powered bombers capable of attaining subsonic and supersonic speeds sometime between 1966 and 1969.[89] After some prompting from the Joint Committee on Atomic Energy of the Congress, the Air Force and the AEC proposed an accelerated nuclear flight program intended to regain the "psychological edge" from the Soviets. Eisenhower rejected the request in February 1958 although he approved continued reactor development.[90]

But the idea of developing an entire weapons system was far from dead within the Air Force. In the spring of 1958, SAC proposed combining nuclear power with the stand-off missile mission for an aircraft it called the "continuously airborne missile launcher," or CAMAL. This system would be virtually immune to the two dangers that degraded SAC bomber capabilities in the missile age: a preemptive Soviet attack and Soviet active defenses. Plans called for CAMAL to be able to remain aloft two to five days, carry two ASMs and a 10,000 lb. bomb, fly at a speed of approximately Mach 1, and enter service by 1966. However, the AEC and the OSD believed CAMAL would add only "marginally" to planned strategic weapons for the 1960s, especially since basic research had yielded no results sufficient to justify formal application of nuclear propulsion to specific aircraft. But they suggested continuing research efforts because of the "growth potential" and for "political and psychological" reasons: "It would be unwise, however, to appraise the future value of nuclear propulsion of aircraft on the basis of present knowledge just as it would have been unwise to assign limits to jet engine performance in the early 1940s. . . . We believe there is no question but that the U.S. public and the world at large will attach great significance to first nuclear flight as positive evidence of the relative technical statures of the United States and the U.S.S.R., regardless of the real military value of the accomplishment."[91] President Eisenhower endorsed this approach in early 1959; he told Defense Secretary McElroy and Deputy Defense Secretary Quarles that "the Government should concentrate on the development of the power plant."[92] The OSD reservations about nuclear-powered flight arose in part because of the tight budget—it simply could not afford to waste funds on projects that would have minimal returns during a period of fiscal stringency.

Pressure by nuclear propulsion advocates in Congress, concentrated in the JCAE, led Quarles to promise a thorough review of the research program, particularly the work done by General Electric. After a visit to GE in May 1959, Quarles evidently decided that the nuclear propulsion program warranted an additional $25 million.[93] But Quarles did not implement his decision, for he died in his sleep the same night. Defense Secretary McElroy directed Quarles's replacement, Thomas Gates, and ARPA Director York to examine the nuclear propulsion program. The nuclear-powered bomber gained some support from the WSEG, particularly for use in ASW, warning, logistic, and strategic bombing missions. According to Robert Little, WSEG Report no. 37 ("Evaluation of Military Applications of Nuclear Powered Aircraft") claimed "CAMAL had a pronounced advantage over both the B-52 and the proposed B-70 if the reactor had an operating life of 1,000 hours." The WSEG endorsed an expanded experimental nuclear-powered flight program.[94] The JCS later approved this approach.

In late June 1959, ARPA Director York completed his review of the nuclear propulsion program and presented it to the president, Gates, AEC Chairman John McCone, Killian, Kistiakowsky, Gray, and Goodpaster. A nuclear-powered aircraft, York told them, had been under development for thirteen years at a total cost of $900 million. Current plans called for an additional expenditure of $400 million over the next four years for development of a 600,000-pound nuclear-powered plane.[95] In York's opinion, continuation of the nuclear propulsion program would cost far more than this amount and the aircraft would not become operational until at least 1970. The ARPA director stated that the primary difficulty—development of a nuclear reactor small enough to fit on a plane and still produce enormous amounts of energy—still had not been overcome. He proposed limiting research to "the reactor-engine combination rather than the other elements of the program." Goodpaster recorded that "the President vehemently agreed, commenting that the only difference he has is with the mild way in which Dr. York put this."[96] The other senior advisers agreed with York to varying degrees but those supporting a more aggressive program refrained from voicing their support because of the poor test results, the dimmer prospects, and the budget situation.

Research on nuclear-propelled aircraft continued on a limited basis during the remainder of the Eisenhower administration. Restrictions on funding prevented the reactor research from ever becoming a high priority. The administration had given up on nuclear-powered flight as a strategic and economic alternative to conventional aircraft but allowed a minimal amount of funding, to placate advocates in Congress.

The nation's strategic bombers underwent significant changes during the missile gap period. Uncertainty about Soviet missile programs and U.S. warning capabilities led to the reconfiguration of SAC peacetime operations through dispersal, ground alert, airborne alert training, and Positive Control. With the exception of the levels of airborne alert, short-term measures to reduce SAC vulnerability met with little resistance or criticism from the administration, the other armed services, or Congress. However, the Air Force's efforts to expand bomber capabilities encountered numerous technical and political problems. The sole exception, the Hound Dog ASM, was successful because it required no major technological advancements and it would expand U.S. strategic capabilities before ICBM deployment. The other three programs—Skybolt, the B-70, and the nuclear-powered bomber—possessed neither virtue. These three weapons offered a tremendous improvement in bomber capability and performance but required engineering feats beyond the state-of-the-art. Air Force leaders erroneously thought that technology could be pushed in developing these weapons, just as it had been in the past. But each development program suffered delays and escalating costs when the technological advances failed to materialize. In other times, civilians might tolerate delays and rising costs to develop a new, revolutionary weapon. But with deployment of strategic missiles expected to be only a short time away, the Eisenhower administration viewed these bomber programs as strategically unnecessary and fiscally extravagant.

President Eisenhower's unwillingness to support bomber development programs initiated a de-emphasis of bombers in favor of strategic missiles. SAC reached its Cold War peak of 1,854 bombers in 1959, but the number began to decline almost immediately because of retirement of the B-47 (which constituted two-thirds of SAC bombers) and the end of B-52 production. Technical problems and the Eisenhower administration's priorities meant there would be no new bomber for the Air Force for many years. In January 1961, the Air Force hoped and expected that the Kennedy administration would keep its campaign promise to build the B-70. But Kennedy and Robert McNamara soon reached the same conclusion reached by Eisenhower: the B-70 was technically questionable and would be simply too expensive after the purchase of hundreds of ICBMs and SLBMs. Air Force leaders, particularly Generals White, LeMay, and Power, had gambled on exotic aircraft to protect manned bombers in the missile age, and had lost.

Strategic Missile Force Planning

In October 1957, the U.S. strategic missile program was still years away from deploying an operational weapon. Decisions over the previous decade had slowed missile progress. They included the Truman administration's insistence on tight defense budgets before the Korean War, the Air Force's choice not to research strategic missiles prior to 1953, the Eisenhower administration's imposition of budget limits, the separation of military and scientific rocket programs, and the approval of multiple military missile programs rather than a concentrated effort. The Eisenhower administration confronted numerous questions concerning missile development and deployment after Sputnik. Missile system development, which raised technical issues and the issue of testing, dominated missile discussions from the launch of Sputnik I until late 1959. After testing had proved the technical feasibility of strategic missile systems, the focus of the administration's deliberations shifted to the composition and size of missile deployments. Administration officials were still confronting questions concerning missile deployments when they left office in January 1961. Their missile decisions were complicated by public pressure to catch up with the Soviets, the president's budget limits, and the development of multiple missile systems (3 ICBM, 2 IRBM, and 1 SLBM), which meant a continuation of the pre-Sputnik problem of allocating limited funds among many systems. Interservice rivalry sharpened this competition between missiles, since each service had at least one ballistic missile program.

Ballistic Missiles from Sputnik I to 1959: Overcoming Technical Obstacles

Sputnik I's launch demonstrated that the United States trailed the Soviet Union in a number of areas, including rocket thrust.[97] In the first days after Sputnik, Eisenhower and his senior advisers sought to formulate a response. Clearly, Eisenhower had underestimated Sputnik's psychological effects, in part because the military benefits of first-generation missiles seemed negligible in comparison to bombers. Some aides suggested merging the scientific and military missile programs but Eisenhower rejected this. He told the NSC that "we have a plan—a good plan—and . . . we are going to stick to it," although he lifted some of the restrictions on missile overtime that had been imposed in the summer of 1957. Eisenhower blamed some of the predicament on the military for not following through on his 1955 directives to develop missiles—apparently oblivious of the effect of his own budget restrictions, imposed only months earlier. On several occasions, he floated the idea of reorganizing military missile programs into another "Manhattan Project," but nothing ever came of it.[98] Thus, in the first weeks after Sputnik, Eisenhower accelerated missile programs but rejected the crash program that some had advocated, largely because of the remaining technical uncertainty concerning missile development.

By early November, the Gaither report had increased pressure within the administration for an expanded ICBM program. The Gaither panel claimed that the Soviets "probably surpassed" the United States in ICBM development and expected that there would be a "critical period" in 1959, following Soviet deployment of a "significant" ICBM force. To prevent a missile gap, it advocated increasing the planned ICBM deployments from 80 to 600, accelerating first-generation ICBMs so that initial deployments could begin in 1959, and accelerating Polaris production and increasing the Polaris programmed force from 6 submarines to 18.[99] The conclusions of the Gaither report were reinforced by SNIE 11-10-57 (issued on December 17), in which U.S. intelligence predicted a rapid Soviet ICBM buildup—an increase from 10 in 1958 to 500 in late 1960 (see Chapter 2). The president approved a variety of measures to accelerate ballistic missile programs without having to initiate crash programs that would demand even greater funding. But he refused to make a rash decision to increase programmed ICBM forces, especially since first-generation missiles had not yet proven their technical viability in flight tests. Instead, Eisenhower decided that each Gaither recommendation would be examined by the DOD and the armed services and that this would be followed by an NSC evaluation, beginning in late January 1958.

As the DOD was reviewing missile programs, James Killian, Eisenhower's first science adviser, investigated the technical progress of each missile system. He formed the Ballistic Missiles Panel of the PSAC in order to provide the president with a continuing evaluation. In its initial report, issued on December 28, 1957, the panel acknowledged that the United States lagged behind the Soviets but claimed that progress "has been impressive." It expressed "confidence" that there would be an early missile deployment and encouraged additional research, to be applied in the development of second-generation missile systems.[100]

The NSC discussed the Gaither recommendations and the DOD analysis at two meetings in January 1958. Deputy Defense Secretary Quarles explained to the NSC that the Navy was formulating a plan to accelerate initial Polaris deployments by two years and to increase the number of submarines being constructed from 3 to 9. He warned that this plan required much more money, but he also "predicted very strong Congressional support for the construction of perhaps as many as 100" Polaris submarines. Quarles noted that the United States had programmed 130 ICBMs (90 Atlas and 40 Titan) for deployment by 1963, not 80 as stated in the Gaither report. However, the U.S. ICBM forces would still be much smaller than the Soviet Union's if intelligence estimates proved correct. According to Quarles, the difficulty in meeting the Gaither recommendation for deployment of 600 ICBMs lay in base construction not actual missile production: "In fact, if we were to have 600 ICBMs operational by FY 1963, we would have to begin the construction of bases at once. By and large, the Department of Defense [thinks] it unwise to undertake this program."[101] Later in the meeting, Eisenhower expressed his "considerable anxiety about the necessity of proceeding to the production of certain ballistic missiles without full testing." He added: "After achieving the production of a certain number of such ballistic missiles—the number deemed absolutely necessary—we should flatten out the production curve until further testing [has] resulted in the perfecting of the missiles in question."[102] Thus, the Eisenhower administration did not object to the size of the ICBM force recommended by the Gaither report but to the deployment of technically deficient or inadequately tested ICBMs.

The Gaither report continued to provide the framework for a lengthy NSC examination of missiles and other defense issues throughout the winter of 1957 and the spring of 1958. Defense Secretary McElroy requested the JCS's views on increasing the planned force level of Polaris to more than 3 submarines and the planned deployment of ICBMs to more than 130 missiles.[103] But the JCS split over missile expansion and failed to meet the NSC deadline of late January. The Air Force and Navy prepared plans for dramatic increases in their strategic missiles programs. The Air Force wanted to more than double the programmed ICBM force, from 130 missiles to 270.[104] Navy Secretary Gates, drawing on the January 6th NSC briefing, recommended increasing the planned Polaris force level to 9 submarines by the end of 1961 with a subsequent capability to produce one submarine per month. The deployment of the first submarine could be achieved by December 1959, although nuclear warheads would not be available until March 1960.[105] Initially, the Polaris missile would deliver a 300–500 KT warhead a distance of 1,100 miles with an expected CEP of 3–4 miles, though by 1963 the range could be increased to 1,500 miles and the CEP reduced to 2 miles. The cost of the rapid accelerations proposed by each service led McElroy to defer any expansion.

The nation's missile research in the months after Sputnik exceeded expectations in such critical areas as solid propellants, inertial guidance, small-weight megaton warheads, and warhead nose cones. These advancements put the initial deployments of first-generation missiles on or ahead of schedule. One major problem that remained was solid fuel performance—a key component in second-generation missiles such as Polaris and the Air Force's new Minuteman ICBM. As PSAC member George Kistiakowsky wrote Killian, "to develop a solid propellant ICBM is a still more difficult undertaking and its earliest availability is 1965, while 1966–67 is a more realistic date, unless a crash program is initiated."[106] The PSAC Ballistic Missiles Panel endorsed improving liquid fuel in light of the solid fuel problem and pressed for a choice between the competing ICBM and IRBM systems.[107] Killian, Kistiakowsky, and York informed Eisenhower of their opinions in early February. Once again, Eisenhower expressed regret at not centralizing missile programs and "stressed that what we must have is the earliest possible operational capability." But he added a qualification: "We should not try to excel in everything. . . . Psychological as well as technical considerations are important—at times appearances are as significant as the reality, if not more so."[108]

The PSAC Ballistic Missiles Panel delivered a longer technical analysis of the retaliatory role of missiles to Killian on March 4, 1958. The panel concluded that first-generation missiles would have "limited effectiveness" for retaliation because of liquid fuel, base complexity and soft configuration, poor guidance, and warhead design. The panel found the Air Force's plan for the smaller second-generation solid-fueled Minuteman ICBM lacking as well, calling it "another crash program" unlikely to be deployed on schedule. It warned that vulnerability and guidance problems might preclude a Minuteman counterforce mission. The panel offered two alternatives to the Air Force's Minuteman program. The first plan would delay Minuteman development for one to two years until better solid fuels could be produced. Titan refinements would compensate for any loss in ICBM capability. Under this plan, the improved Titan would be operational in 1963 and Minuteman in 1965; the Atlas would be discontinued when feasible. The panel argued that "this mixed force may offer considerable operational advantages. U.S.S.R. might be forced to devise its active missile defense as if all our weapons had counter-measure capabilities of the Titan nose cones. At the same time, it might have to provide for the large numbers of our weapons which the low cost of Minuteman would permit us to deploy." The second plan would combine Titan modifications, increased solid fuel research, and Minuteman deployments on the periphery of North America as a hedge against inadequate propulsion. The panel favored either alternative plan over the Air Force's plan, though it clearly leaned toward the first alternative.[109]

Killian and Kistiakowsky discussed the panel's report with Eisenhower on March 10. Killian emphasized the need to review the entire missile program before making decisions on force size, configuration, basing, and planning.[110] Eisenhower agreed and approved Killian's recommendations for making Titan improvements and for consolidating solid fuels research under the auspices of the DOD's new Advanced Research Projects Agency. The president also agreed with their analysis of Minuteman but deferred a decision pending receipt of further reports and studies.[111]

Killian had provided Eisenhower with ample background for evaluating the DOD's request for supplemental missile funding. In a meeting with McElroy and Quarles on March 20, the president pressed them for details on many of the issues raised in the panel's reports: the pace of Minuteman development, ARPA control of solid fuel research, and greater centralization of missile development. Even though he approved the DOD accelerations, Eisenhower remained skeptical of second-generation programs and worried about unnecessary overlap between Polaris and Minuteman.[112]

The Defense Department finally presented its report on missile programs to the NSC on April 24, 1958—almost three months later than originally scheduled by Culter. Director of Guided Missiles Holaday briefed the NSC on plans for Polaris and ICBMs, as well as other issues.

The DOD had rejected the armed services' plans and recommended no changes in programmed ICBM or SLBM force levels. Still, the president and others expressed their unease about various aspects of the missile programs. Eisenhower and John Foster Dulles worried about the rising costs of missiles. Killian asserted that until the WSEG completed a report on strategic weapons, "we are obliged to resort to *ad hoc* decisions such as had been done in the case of deploying IRBMs in the NATO area." Eisenhower also questioned the simultaneous development of first- and second-generation missiles: "We are now beginning to think of aircraft as becoming obsolescent, and so it is also with first-generation ballistic missiles. Despite this, we are going ahead full steam on the production of both aircraft and first-generation ballistic missiles."[113] Conscious of the budgetary implications of developing multiple missile systems simultaneously, the administration was determined to move cautiously until the missiles had been proven technologically viable.

Congress tried to accelerate Polaris in June 1958 by adding $638 million to the administration's DOD supplemental appropriation—in effect funding the nine-submarine program proposed by the Navy in January—which would increase Polaris expenditures to $2 billion for FYs 1959 and 1960.[114] Budget Director Stans told Eisenhower that the administration's request was "sufficient." The increase prompted the president to tell CJCS Twining that "we are apparently planning to 'kill every Russian three times' in the development of our forces for massive retaliatory attack."[115] In July, Eisenhower approved initiating construction of the fourth and fifth Polaris SSBNs but decided to delay spending the money that had been appropriated for the other four submarines.[116] This occurred even before the Navy had conducted a missile flight test.

Two aspects of Polaris development worried the PSAC Ballistic Missiles Panel. First, inertial navigation for the missile showed such little promise that the Navy contracted North American Aviation to develop an alternative system for the initial deployments. Second, missile weight and solid fuel inefficiency meant that the range of Polaris would be less than 1,000 NM rather than the goal of 1,500 NM. The panel concluded that "the ultimate Polaris with 1,500 n. mile range may become operational only in a somewhat more distant future than the planned 1963."[117]

By mid-1958, the United States was at the threshold of the missile age. The Eisenhower administration had set initial force levels and had already increased them, had budgeted funds for the purchase of some missiles, and had moved quickly to develop second-generation missiles. But the administration still faced considerable uncertainty. Each strategic missile program (Atlas, Titan, Minuteman, and Polaris) confronted technical obstacles in development as well as certain operational liabilities—due in large part to the unfamiliarity with the performance of these technologies. This, of course, complicated the administration's efforts to build a missile force that would have military utility, not just a missile "Potemkin village." Moreover, budget restrictions denied the administration the luxury of delaying the choice between competing ballistic missile systems.[118]

Atlas, the ICBM closest to deployment, continued on a "very tight" development schedule to ensure initial deployments in mid-1959. The first Atlas flight test took place in July 1958; the second, in which the missile which traveled a full intercontinental range of 5,500 miles, followed in November.[119] The Air Force also modified Atlas deployment plans to include greater dispersal, inertial guidance, and the semihardening of bases to 25 psi after the first thirty deployments. Despite these changes, doubts persisted about Atlas's long-term military utility. The ICBM's liquid fuel propulsion system required many preparations before firing. This, combined with the overall complexity of the system, meant that Atlas would not be a very responsive weapon, whether deployed in a "soft" or in a semihardened configuration. According to one of the developers, Atlas's "operational complexity will be such that it will not be considered satisfactory by the Air Force."[120]

The Air Force's concern about the military effectiveness of missiles led to further Titan modifications in 1958. The Air Force researched two storable liquid propellants that would improve Titan operational reliability by facilitating shorter, less complicated firing procedures and more efficient flights. Killian approved of the propellant conversion program but called it "a very large task" in view of the cost ($350 million to $400 million) and because it probably would not be available until after completion of the first Titan deployments. The Air Force also developed plans for hardening Titan to 100 psi to decrease vulnerability. But hardened bases cost $25 million more to construct than soft bases (and $14 million more than semihardened ones) and required fifty-six months to build. Consequently, Titan's operational modifications demanded more money immediately even though the missile's only flight test had failed.[121]

Titan's rising costs and delays in development prompted consideration of its cancellation. With Atlas needed for an early missile capability and Minuteman needed because of its technological advancements,[122] tight budgets and Eisenhower's pressure to reduce the overlap in strategic weapons seemed to make Titan the "odd missile out." In August 1958, in preparation for his FY 1960 budget recommendation, Defense Secretary McElroy asked the Air Force Ballistic Missile Division (BMD) for its recommendation concerning expansion of programmed ICBMs to 200 from 130 (90 Atlas and 40 Titan). The BMD compared two force postures: 200 Atlas, and a mixed force of 110 Titan and 90 Atlas. On the surface, the result of canceling Titan in favor of more Atlas missiles would be a force with nearly identical capabilities and a saving of more than $400 million over the next five years. The BMD recommended the mixed ICBM force, however, because of Titan's potential for improvements in range, propulsion, guidance, and reliability. But Titan cancellation appealed to some in the Air Force and to DOD Director of Guided Missiles Holaday. Throughout the fall, the head of the Ballistic Missile Division, General Bernard Schriever, lobbied Washington policymakers in an effort to save Titan. The BMD thought that evaluations of missiles "are best appreciated by the people closest to the problem, and consequently are not likely to be weighed properly by people making the decisions."[123] Schriever aimed his criticism at Holaday, referring to his "broken promises, double dealing, and deliberate delays."[124] His efforts paid off when McElroy approved the mixed ICBM force on November 13 and included it in the defense budget for FY 1960.[125]

The administration's missile decisions in late 1958 were also influenced by two reports by the Weapons Systems Evaluation Group on strategic weapons: WSEG-23/1 and WSEG-30. In the first, the WSEG argued, as it had in an earlier report, that problems of reliability and inaccuracy limited the military utility of first-generation ICBMs. It calculated that 130 ICBMs and 80 SLBMs as well as an additional 30 IRBMs would be needed to destroy only 50 soft targets. Obviously, the role of any first-generation missile in a counterforce attack would be negligible. WSEG-23/1 encouraged the emphasis on second-generation missiles, since they would be three times more survivable as a result of dispersal and hardening.[126] The ICBMs could play a major deterrent role only in 1962—after sizable deployments.[127] The second report, WSEG-30, reinforced these conclusions. It indicated that Titan and Polaris would be the most cost-effective and the most survivable weapons at the time when the Soviet ICBM lead was anticipated to be greatest (1960–1962). WSEG-30 warned that decisions about increasing the two systems would have to be made "very soon" because of the length of time required for base construction. It emphasized the value of Minuteman in any missile force: "Even in the face of the most severe attack by Soviet missiles, the exchange rate in terms of numbers would be very favorable for the U.S. One Minuteman squadron of 100 missiles . . . would severely tax the estimated Soviet production capabilities if they were to attemp [*sic*] to produce enough ICBM's to permit neutralizing a Minuteman squadron."[128] The two WSEG reports effectively made the case for delaying large ICBM deployments until second-generation missiles became available because of their increased technical effectiveness.

Both WSEG reports were circulated throughout the administration, although the NSC was briefed only on WSEG-30. President Eisenhower praised that report but criticized it for failing to indicate which strategic weapons could be eliminated. He and his special assistant for national security affairs, Gordon Gray, pressed the JCS for advice as to which strategic forces the United States should possess (see Chapter 4).[129] Despite the effort by the White House to contain missile costs by eliminating redundant strategic weapons, McElroy recommended increasing the number of Polaris submarines from 6 to 12 and endorsed the mixed 200 ICBM force. As discussed in Chapter 4, this proposal aroused the concern of the president, but he eventually approved the budget recommendation.

By the end of 1958, the strategic missile force for the following decade was beginning to come into focus. First-generation missiles would be deployed only as part of a token force to attain a symbolic operational missile capability. The administration would wait, much to the consternation of missile gap critics such as Symington, for technical improvements in second-generation missiles before making large deployments.

Missile Deployment Decisions, 1959–1960

The Eisenhower administration's strategic missile decisions in 1959 and 1960 revolved around missile deployments, particularly the second-generation Minuteman and Polaris missiles. While large deployments seemed virtually certain in early 1959, the size, pace, and composition of the missile buildup remained unknown. The answers to such questions meant billions of dollars to the services. They provoked a bitter and intense debate between the Air Force and the Navy that spilled over into other policy areas such as nuclear strategy, targeting authority, and, of course, budgets (see Chapters 3 and 4).

The technical progress and strategic value of Minuteman led the administration to place greater emphasis on the weapon in the first half of 1959. The JCS determined in February 1959 that "an early IOC is important and . . . Minuteman is required as soon as possible without a crash program."[130] This recommendation came as the administration defended its defense budget in Congress while being urged to accelerate the purchase of ICBMs.[131] McElroy and Quarles decided to propose adding Minuteman to the NSC list of "highest priority" weapons programs.[132] But McElroy decided in May to delay increasing Minuteman's priority even though he "was now convinced that Minuteman should" receive it soon.[133] By August, McElroy and Kistiakowsky "felt it important to give an unlimited priority now to Minuteman" and received Eisenhower's approval to add it to the NSC list.[134]

The Air Force developed a MINUTEMAN deployment plan as part of its budgeting and force planning exercises in spring 1959. The plan called for a rapid and astronomical deployment of MINUTEMAN throughout the 1960s, as shown in the table.

Air Force MINUTEMAN deployment schedule, 1962–1968

Year	MINUTEMAN force size
1962	0
1963	150
1964	650
1965	1,350
1966	2,050
1967	2,800
1968	3,000

Deployments would wind down only when a new advanced ICBM was beginning to be deployed.[135] The deployment projections in this plan vastly exceeded the informal deployment projections circulating within the administration, probably because the Air Force wanted to capitalize on the missile gap debate then raging in Congress.[136]

With budget preparations under way and Minuteman production lines due to open, the OSD sought the advice of the JCS as to the initial Minuteman force levels. After a JCS split, a report by the Joint Staff endorsed Minuteman force levels for the first year ranging from 100 missiles to 150 missiles. The Joint Staff recommended against establishing force levels beyond this range because of the many uncertainties that surrounded defense planning, including the BNSP, the size of the target system, the availability of nuclear matériel, and the resources available for defense programs. None of the armed services agreed with these conclusions. The Air Force tried to protect the first step of its ten-year Minuteman plan. It argued for an initial deployment of 150 Minuteman missiles rather than a general range because "a force objective of at least 150 operational Minuteman missiles by end FY 1963 is considerably below the anticipated total requirement and represents a modest initial step in achieving an adequate capability with improved solid propellant ICBM's." The Joint Staff opposed the Air Force position on the ground that a clearly defined objective "tends to predetermine the forces." In a joint statement, the Army and Navy asserted that initial deployments should not exceed 50 Minuteman missiles until the WSEG had completed an ongoing study on strategic weapons. The Joint Staff rejected this advice as well. After yet another JCS split, the OSD approved an initial force size of 150 Minuteman missiles; presidential approval would not be officially received for a number of months, however.[137]

One feature which distinguished Minuteman from other ICBMs was that plans called for it to be deployed in a mobile configuration as well as in hardened silos. The mobile mode required placing Minuteman missiles on railroad cars and shuttling them across the country. The Air Force hoped a mobile ICBM would be a hedge against ICBM vulnerability once the Soviets had begun to deploy accurate ICBMs.[138] Equally important, a mobile ICBM would steal the Navy's most effective argument in favor of Polaris (invulnerability through mobility), thus aiding the Air Force in budget battles over missiles. Early Air Force plans called for 10 percent of all Minuteman missiles to be deployed in the mobile mode. According to the plans of General Motors, each train would have seven cars, as well as another two to eight cars with one Minuteman ICBM per car.[139] Air Force Chief of Staff White supported the idea of a mobile Minuteman and urged the rest of the Air Force to do so. He told Vice Chief of Staff LeMay, "we should get *Mobile* Minuteman *fast*. I want a larger proportion of Minuteman in mobile configuration with max feasible number of missiles per train."[140] In late November, the Air Force and SAC broadened the weapon's appeal to the administration by reducing the costs of the mobile Minuteman; to this end, they increased the number of missiles per train from three to five and implemented less stringent guidance requirements.[141] SAC planned to deploy 500 Minuteman missiles on 100 trains across the country by 1964.[142] Defense Secretary McElroy approved, at least in a general sense, of the increased emphasis on mobility in missiles.[143]

The Eisenhower administration encountered an entirely different set of problems with regard to the Atlas and Titan programs throughout 1959. Each missile had had some disappointing test results, which prompted revisions in the development and deployment schedules.[144] However, these technical missteps troubled the administration less than did the escalating costs of base construction and the Titan program's managerial problems. First-generation missiles required greater support because of their design. This, in addition the uniqueness of rocketry, translated into very complicated base designs. "The complexity of earliest missiles bases," Kistiakowsky told Eisenhower, "will be almost beyond imagination."[145] The modification of deployment plans to account for hardening, storable fuel, and quicker reaction time increased base complexity and resulted skyrocketing in base construction costs. McElroy wrote that "the information that has been made available to me indicates that the construction costs alone have almost doubled" because of hardening and dispersal.[146] The administration estimated the costs of deploying nine Titan missiles to be $122 million—$102 million for construction of the bases alone.[147] President Eisenhower expressed an interest in reducing the costs of missiles, even if it meant increasing reaction time. He warned of the effects of unrestrained weapon programs and "cautioned against a situation whereby we tend to take the sum total of everybody's optimum requirements and thereby break ourselves."[148] He also considered a comprehensive reorganization of the ballistic missile program and called for a study of it.[149]

Titan still developed fitfully in 1959 because of management problems of the prime contractor, the Martin Company. Testing had validated the decision against termination a year earlier, but Martin's management of Titan had been causing problems for the Air Force for three years.[150] When the Air Force examined the test delays in 1959, it found that most "were directly attributable to personnel errors which indicates the need for a greater tightening of discipline and management procedures within the company." Schriever, now head of the Air Research and Development Command, "demanded" that Martin "replace its management personnel with the most skilled executives" available and refused to award contracts to Martin for other Air Force projects.[151] The scientific community supported the Air Force's actions; Kistiakowsky, summing up the conventional wisdom, said that Titan "has good design and engineering . . . but . . . it is a management mess." However, Titan's engineering advances compensated for Martin's inefficiency and helped to keep the program close to the development schedule.[152]

Problems of base construction and Titan management had temporarily insulated the Eisenhower administration from Air Force demands for increased ICBM force levels. Prompted by the missile gap debate in the spring of 1959, the Air Force recommended to McElroy that the programmed ICBM force level be increased from 200 missiles to 290 (170 Atlas and 120 Titan). This increase would cost $365 million immediately and another $3.7 billion over the next three years. But McElroy balked and the idea died after Congress failed to appropriate enough money to fund the increase. Several times in the next few months, the Air Force recommended increasing ICBM force levels to 290 missiles, but to no avail.[153] But during the FY 1961 budget deliberations in November and December 1959, the question of missile deployment levels was raised once again. By this time there was an impetus to move beyond the NSC approved force levels of 200 ICBMs and 9 Polaris submarines—most notably, McElroy's approval (in August) of an initial force of 150 Minuteman missiles and congressional appropriation of funds for a Polaris force of more than 9 submarines. Cognizant of the budget limits and the redundancy in missile systems, McElroy sought only a modest increase in programmed force levels in the budget for FY 1961. In November 1959, he proposed increasing the ICBM force level to 270 missiles (140 Titan and 130 Atlas) and increasing that of Polaris to 15 submarines (12 fully funded and 3 with start-up funds only).[154] The president approved McElroy's recommendation with the exception of the Titan increase, most likely because of the continuing management problems. With the B-70 dominating budget deliberation, these missile decisions were accepted without much discussion.[155]

The Eisenhower administration's measured increase in missile force levels frustrated the Air Force and the Navy. The Navy projected a

Polaris force of 45 submarines carrying 720 SLBMs; in the spring of 1959, the Air Force planned for an ICBM force of 170 Atlas, 120 Titan, and 3,000 Minuteman missiles.[156] Most of the service's uncertainty arose from Minuteman. The Minuteman force size recommended by the administration would affect procurement decisions concerning other missile programs as well as bombers. But McElroy decided not to include Minuteman force levels in the budget for FY 1961 and left the issue to his successor, Thomas Gates. In the previous three years, the administration had resisted any commitment to large ICBM deployments on the grounds that first-generation missiles had only marginal military utility. Now, at the beginning of 1960, an ICBM that possessed military advantages awaited a presidential decision.

The administration's consideration of Minuteman in early 1960 focused on both technical issues and force levels. On the technical side, deploying Minuteman in underground silos for long periods presented numerous new hurdles. The missile, according to Kistiakowsky, required "development of a miniaturized inertial guidance system which will not only be accurate enough but will remain so after years of sitting in the hardened sites, with the gyroscopes running all the time." Technical obstacles precluded accelerating Minuteman, even though they would "eventually" be overcome.[157] Eisenhower seemed favorably disposed toward Minuteman; he told the NSC in February that "as soon as our more advanced missiles [prove] themselves, we must see what we can do to get them into production. [I do] not wish to create a big arsenal of weapons which [are] not yet fully developed or weapons which [will] soon be obsolete."[158]

With the White House supportive of Minuteman, Deputy Defense Secretary James Douglas requested on April 1, 1960, that the NSC approve production and initial deployment of 150 Minuteman missiles by mid-1963.[159] He told the NSC: "If successful, [Minuteman] would close the so-called 'missile gap.'" Eisenhower replied that "perhaps we should go ahead with this program if the scientists [are] convinced that Minuteman is an operational weapon." However, "in the absence of tests [I am] slightly skeptical; [I'd hate] to buy 'a pig in a poke.'" But Douglas assured that president that this plan was necessary so that deployments could begin in 1962. When Eisenhower asked about the initial level of Minuteman production, Douglas explained that "there [is] no target beyond 400 missiles by the end of . . . 1963," although that level would require "substantial funding" in the next budget. This open-ended approach to Minuteman production rankled Eisenhower. He admonished Douglas: "We [are] inclined to say what the final figure for production of a missile would be as soon as we start producing it." Douglas claimed he had been "conservative" in making the production decision and had fought off Air Force attempts to set production at 800 Minuteman missiles by 1964. An annoyed Eisenhower retorted: "Perhaps we should go crazy and produce 10,000 Minuteman." The NSC officials discussed the timing of U.S. missile deployments with Douglas, noting that by 1963 the United States would have a larger force than the Soviets, according to the most recent NIE. Eisenhower approved Douglas's request, although with some resignation: "we [have] gambled so much on our missile program since 1955 that we might as well take another gamble. . . . [I hope] the Department of Defense [will] be as eloquent in suggesting the abandonment of unnecessary weapons systems as it [has] been in proposing the commitment of Minuteman to production."[160] Eisenhower and his aides restricted the progress of first-generation missile programs throughout 1958 and 1959, citing the numerous operational advantages of second-generation missiles. But Eisenhower's enthusiasm for second-generation missiles dissipated when he confronted Minuteman's scope and costs.

The PSAC Missiles Panel, in response to Eisenhower's concerns, completed a comprehensive review of the Minuteman program within a month after the NSC meeting. The panel's report emphasized Minuteman's strategic value and the desirability of early deployments: "The Minuteman will do two important jobs in the period 1963–6; one is the creation at the lowest cost of a force which could be destroyed by the Soviets only if they were to build a large number of missiles of great accuracy. The second job is that of supplying a basically different component to our 'mix' of missiles, i.e., the land based rail mobile system." But the panel noted various technical problems, particularly those arising from the requirement for a thirty-second response time. Additionally, it criticized the Air Force for not providing a retargeting capability for the fixed Minuteman missiles: "According to current plans, missiles in fixed sites will be pre-set to go to only a single target. It would probably be highly desirable to have each missile pre-set to fire on command at any one of several targets. This could reduce the amount of 'overkill' that would otherwise need to be built into a retaliatory firing plan, and thus would reduce the total number of missiles required." Aware of Eisenhower's concerns about the cost of multiple missile programs, the PSAC panel advised that Minuteman and Polaris be given equal priority and that Skybolt should have a secondary emphasis. Further, it asserted that the balance between fixed and mobile Minuteman missiles should be determined by the missile balance existing between the United States and the Soviet Union: "We are inclined to believe that the hard and dispersed Minuteman might merit higher priority than the mobile system, though this should probably be true only if we were willing to build a force at least half as large as what is believed to be a reasonable upper limit on the size of the Soviet ICBM force. . . . In the event of a decision in favor of a smaller force, the conclusion on the relative priorities would be reversed with the mobile system then meriting primary emphasis."[161] No "combination of ballistic missiles," the panel concluded, "can assure the U.S. a secure retaliatory posture during the next 12 to 18 months." If Eisenhower wanted to find a reason to restrain the Minuteman program, the PSAC panel's report failed to provide one.

Presidential science adviser Kistiakowsky and Franklin Long of the Missiles Panel briefed Eisenhower and several White House aides on the panel's report on May 4.[162] Long informed the president that Air Force firing policy called for volleys of fifty missiles and that a volley could not be interrupted once initiated. Eisenhower said "we allow for no margin of error, and raise the chances of starting a war that no one wanted. . . . [I]t is better to take a few extra minutes, to give someone high up in authority the decision." He approved an examination of a retargeting capability for Minuteman and concurred with the PSAC panel's assessment favoring fixed over mobile Minuteman deployments, but endorsed "having at least a few of the mobile type, since in these matters it is best to have every kind of string on one's bow."[163]

Despite his concerns about budgets, strategic and operational factors won over Eisenhower in this round of Minuteman decisions. He approved the production and initial deployment of 150 Minuteman missiles over the next three years. The president also knew that plans called for 805 operational Minuteman missiles (691 fixed and 114 mobile) by mid-1964. Considering the other 270 programmed ICBMs, the 240 SLBMs, and the bomber force, it becomes clear that Eisenhower had taken great strides toward expanding America's strategic arsenal to even higher levels of overkill.

While the administration was evaluating the Minuteman programs, there was pressure for increasing the Polaris program. Flight tests in 1959 and 1960 were very successful, with the "first fully guided" missile landing 0.1 NM from its target after traveling 900 miles.[164] Despite his enthusiasm for Polaris, Eisenhower had initially refrained from moving beyond the fifteen-submarine program until the Navy had tested the "cold launch" technique.[165] But Defense Secretary Gates advocated an increase in Polaris force levels, through a revised budget submission to Congress and transfer of funds from other weapons programs. On April 6, 1960, Gates, CJCS Twining, and Deputy Secretary Douglas met with Eisenhower and his aides to determine the extent of the acceleration. Gates recommended increasing the number of Polaris submarines being constructed from 3 to 5 and increasing the number of long-lead-time submarine procurements from 3 to 6. Under this so-called 5 by 6 plan, the United States would commit funds for the construction of 14 submarines and would begin purchasing the complex components (such as nuclear reactors) of another 6—bringing the total to 20 Polaris submarines. Eisenhower wanted a "3 by 9" plan that would keep the number of fully funded submarines at 12 but would provide long-lead-time funding for another 9.[166] Gates, worried about the possibility of a missile gap in 1963,[167] interjected that under this plan, "we will not get additional boats in 1963, but will get eight in 1964." Eisenhower said that he wished "to avoid a full commitment now" and gently chided Gates for his preoccupation with 1963.[168] The president also rebuffed Gates's request to fund a program to extend the range of Polaris SLBMs from 1,500 NM to 2,500 NM by developing the new A-3 missile. Gates and Douglas noted the program's strategic benefits and the lack of Navy funds necessary to initiate the extension, but the president held firm, claiming that the Navy required no new funding to begin the project.

With Minuteman and Polaris vying for higher priority and funding, rivalry between the Air Force and the Navy became especially intense. Each service feared that loss of a major role in strategic missions would prevent it from competing for large portions of the defense budget, thus possibly undermining the service's long-term stability. The Army meager budgets in the 1950s illustrated what could happen to services that contributed minimally to massive retaliation. As a result, the Navy and the Air Force tried to gain some leverage, particularly in debates about nuclear strategy and targeting procedures (see Chapter 3). The services also sought support for their second-generation systems from of members of Congress, defense intellectuals, and the general public.[169] They circulated within the administration studies that showed the advantage of their system. An Air Force study indicated that 600 Polaris SLBMs would be required to retaliate against 200 targets, whereas the number of Minuteman missiles would vary with the magnitude of the Soviet attack. The Air Force claimed that if the Soviets were to attack with a force of 1,000 ICBMs, retaliating with Polaris would cost 243 percent more than retaliating with Minuteman.[170] A Bureau of the Budget review indicated that, although the Air Force consistently underestimated the costs of Minuteman, Polaris cost from three to six times more than Minuteman.[171]

Events in the spring and early summer of 1960, particularly the U-2 incident, caused a deterioration of superpower relations. In these months Kennedy continued to attack Eisenhower's defense policies, and the administration received the unwelcome news that Atlas deployments had fallen behind schedule. Anticipating that the Soviets might use the presidential campaign to their advantage, Defense Secretary Gates and the JCS presented to the NSC on July 25 a program to increase strategic readiness. The president set the tone early in the discussion when he told the NSC that "we should not let it be known that there is no doubt of the firmness of the present government in doing what is necessary for the security of the country." Gates proposed a larger airborne alert training program, a delay in some B-47 retirements, and increases in conventional forces. The entire program would cost $1 billion, although reprogramming would reduce the amount of new funds needed. Eisenhower "suggested" that the Polaris force be increased by two submarines per year—in effect resurrecting most of the 5 by 6 program he had rejected three months earlier. He said that "we [are] trying to take a cold war action which [will] make our people calmer and the enemy more respectful." When Gates tried to argue for the conventional force increases, Eisenhower answered that "it would be significant if we could say that we were laying down four Polaris submarines a year instead of two."[172] Eisenhower's Polaris expansion bewildered Gates, who said he "believed that we had stepped the Polaris program up to the maximum extent last April." After some NSC members argued again for the conventional force increases, the president staunchly defended the increase in Polaris:

> The President said the only hostilities the U.S. was really concerned about was an all-out atomic attack. He believed that we should be taking military actions which would convince the American people and the Soviet bloc that our retaliatory power has been sharpened and speeded up. He was in

> favor of army modernization but felt the U.S. need not be afraid of brush-fire wars. We could say that we have fought such wars before and could do it again. We should be more concerned about the possibility of a rain of missiles on the U.S. and about becoming so weak that the enemy can attack us with impunity. Such a situation of weakness would affect the mental attitude of both the U.S. and the U.S.S.R. We would not have become so concerned about this matter at the present time except for recent Soviet threats.[173]

Eisenhower's forceful statement of defense priorities effectively ended NSC discussion of Gates's proposal. A few days later, the president officially approved a 5 by 5 program that allowed for the earlier deployment of two submarines as well as the expenditure of $140 million for development of the Polaris A-3 missile and the B-47 proposal.[174]

Eisenhower's determination to increase strategic forces at the expense of conventional forces—in effect, maintaining his New Look approach—continued in the defense budget for FY 1962. The president adopted DOD's missile recommendations with little or no comment. The budget continued funding for the 5 by 5 Polaris program, which increased the force level to 24 submarines (19 fully funded and 5 partially funded) and continued development of the A-3 missile.[175] The administration added 290 Minuteman missiles, increasing the programmed force level to 540 missiles (450 fixed and 90 mobile) by the end of 1964.[176] This could be achieved without increasing production beyond the then current rate of 30 Minuteman missiles per month. In making these decisions, the Eisenhower administration had programmed a massive force of 1,194 strategic missiles: 540 Minuteman, 130 Atlas, 140 Titan, and 384 SLBMs on 24 Polaris submarines.[177]

Shortly after the administration finished its budget, the WSEG reported (in WSEG-50) that ballistic missiles would be the optimum weapon against Soviet targets. With regard to missile systems, the WSEG explained that the balance between Minuteman and Polaris deployments should be contingent upon the Soviet force posture and the scale of the United States retaliatory response. Although the group refrained from endorsing specific force levels, it had determined that mobile systems should be emphasized so long as U.S. deployments remained low. More vulnerable fixed missiles could be substituted for mobile missiles as the U.S. deployments increased. The ICBM force should be weighted in favor of the Minuteman rather than the advanced version of Titan except in instances when special payloads would be needed to overcome active defenses. The WSEG endorsed Polaris A-3 missile development because it would "threaten a larger number of Soviet targets from presently planned deployment areas and . . . will carry the cluster warhead primarily for penetrating possible anti-missile defenses."[178]

Privately, some WSEG members willingly offered more specific recommendations concerning the nation's strategic posture.[179] They argued that a triad, consisting of 300 mobile Minuteman missiles, 20 Polaris submarines, and one-eighth of the total B-52 force armed with Hound Dog missiles on airborne alert, could maintain deterrence by threatening the Soviet Union with intolerable damage. This would be possible, they claimed, in spite of any future countermeasures.[180] Such a recommendation from the WSEG would have been unpopular with both the Air Force and the Navy, since it would justify curtailing almost all the strategic programs then under way. It would also move nuclear strategy in the direction of assured destruction and would preclude any capability to strike first. These ideas were still controversial in the early 1960s, as Robert McNamara would discover.

The U.S. strategic missile program made rapid progress in the three years after Sputnik. The uncertainty about the technological viability of missile programs that had existed early in the missile gap period dissipated as Atlas, Titan, and Polaris moved through the development phase and toward deployment. Additionally, the more advanced Minuteman program, begun immediately after Sputnik, was so successful that the Eisenhower administration set initial force levels only two years into the program. The administration also made important decisions about the pace and scope of the strategic missile buildup. The design and reliability of first-generation ICBMs meant they could provide only marginal strategic benefits over a short period, yet still cost large sums. The administration's decision to restrict these programs at the height of the missile gap debate cost the president politically, as discussed in Chapter 4. But rather than rejecting large missile deployments outright, Eisenhower waited until the second-generation Polaris and Minuteman systems became available before committing to large force levels. Reduced vulnerability, increased reliability, and generally superior engineering ultimately made these missiles militarily useful weapons. Consequently, before leaving office the administration increased the force levels of second-generation missiles to 384 Polaris

SLBMs and 540 Minuteman ICBMs. Most of this increase (15 Polaris submarines carrying 240 SLBMs and all 540 Minuteman ICBMs) occurred in the administration's last fifteen months. Ironically, the administration had finally initiated the buildup that the missile gap critics had clamored for—and did it just as intelligence estimates of Soviet missiles were being revised downward.

The Eisenhower administration's decisions during the missile gap period transformed the nation's strategic forces and established the foundation of U.S. nuclear policy. New Look decisions in the pre-Sputnik era had established the precedent for a large nuclear stockpile. But the potential for an increase in bomber vulnerability as a result of Soviet missile deployments demanded changes in the configuration and composition of U.S. strategic forces. The Eisenhower administration thus altered bomber operations to reduce vulnerability and improved retaliatory capacity by establishing alert programs, Positive Control, and deployment of the Hound Dog ASM. It also diversified strategic forces by developing new strategic missiles and programming very high force levels. In making such missile decisions, the administration emphasized the Minuteman and Polaris systems, which offered more reliability and less vulnerability. These two advantages, as well as others, gave the Eisenhower administration confidence that the United States would possess a secure second-strike capability for years to come. Ironically, Eisenhower's decisions and the Soviet Union's inability to develop a militarily useful ICBM accorded the United States a first-strike capability in the early 1960s. Finally, the Eisenhower administration shaped the U.S. force posture through important decisions about which weapons not to develop or to deploy in significant numbers. Most notably, it began a shift in emphasis, from manned strategic bombers to strategic missiles. As a result of the force planning decisions made by Eisenhower during the missile gap period, the "modern" strategic force began to emerge: a triad of bombers, ICBMs, and SLBMs, each "leg" deployed—or planned to be—in large numbers and configured to reduce vulnerability to a surprise attack.

This examination of strategic force planning during the missile gap period allows us to view President Eisenhower in a new light. Few scholars have noted his role in the strategic missile buildup of the early 1960s,[181] largely because most of the actual deployments occurred after the administration left office and because final decisions about the extent of the buildup were left for the Kennedy administration. Consequently, many revisionist scholars have focused on Eisenhower's attempts to restrain defense spending, gain control over nuclear targeting, or his skepticism with regard to intelligence estimates of Soviet missile programs. But concentrating on such issues without acknowledging the missile expansion results in an incomplete picture. For although Eisenhower restrained defense spending, he also authorized funding for the development and deployment of new strategic missile systems. As he tried to gain more control over nuclear targeting, the president added thousands of strategic weapons and megatons to U.S. war plans. Even after NIEs began to indicate a slower Soviet missile buildup, Eisenhower approved plans for the addition of hundreds of Minuteman and Polaris missiles. If, as Stephen Ambrose argues, Eisenhower's resistance to Sputnik I and the Gaither panel's recommendations for an increase in defense spending constituted "his finest gift to the nation," then the large, secure, and redundant strategic nuclear triad may be his most enduring legacy.[182]

[6]

Eisenhower and the Missile Gap in Perspective

On January 20, 1961, John F. Kennedy succeeded Dwight D. Eisenhower as president of the United States. The new administration moved quickly to carry out its campaign promises to restore the nation's defense capabilities. Administration officials soon found that the situation was not nearly as dire as they had imagined or had claimed. They discovered that there was in fact no missile gap and that if one were to exist, it would most likely favor the United States. Nevertheless, Kennedy and his advisers replaced Eisenhower's budget-conscious New Look policy with a flexible response strategy in the belief that it would restore U.S. credibility by ensuring the capability to respond to a wider range of aggression. The Kennedy administration's strategic force posture represented a continuation of Eisenhower's basic program, although a few programs were accelerated. The administration's major force planning decision concerned when to stop deploying strategic missiles. Defense Secretary Robert McNamara searched for a new nuclear strategy that would be credible and politically viable. Having initially settled on a "controlled response" doctrine, he eventually adopted a declaratory strategy of "assured destruction," although operational strategy continued to be counterforce-oriented. By the time Kennedy was assassinated in November 1963, the nation's defense and nuclear policies had been altered considerably from the New Look. But the choices made by Eisenhower during the missile gap period continued to affect nuclear policy in many ways.[1]

The missile gap period constitutes an important period in the evolution of nuclear strategy, force posture, and ideas. The vulnerability of

U.S. forces to a Soviet surprise attack had been a constant source of concern since the Soviets first detonated an atomic bomb, but now policy makers made decisions with the presumption of vulnerability. Although a missile gap did not materialize as predicted, the Eisenhower administration's choices for coping with the missile age set the course of U.S. strategic policy in an era of mutual vulnerability. These choices represented an important step in the transition of U.S. strategic nuclear policy, from superiority to vulnerability. Recently declassified documents have enabled us to gain new insights into the making of strategic policy during this period—thus helping us to understand the policy process, the bargaining dynamics within it, and the rationales behind the decisions.

This chapter contrasts observations about strategic policy, the Eisenhower administration, and presidential power drawn from this case with the relevant literatures. First it examines Eisenhower's successes and failures in making nuclear policy and compares this assessment with the one popularized by some Eisenhower revisionists. Second, the chapter discusses the contributions of the literatures on the presidency and Eisenhower for understanding this case.

An Assessment of Eisenhower's Nuclear Policies

Eisenhower's most concrete legacy from the missile gap period is the transformation of U.S. strategic nuclear forces. A strategic nuclear triad began to emerge when the Eisenhower administration developed and began to deploy ICBMs and SLBMs. Adding these two missile "legs" to SAC bombers would strengthen deterrence, the administration reasoned, by complicating Soviet plans for launching a first strike regardless of the vulnerabilities that might arise for any particular leg. Because he viewed military effectiveness and invulnerability as more important than early deployment, Eisenhower approved only a token deployment of first-generation missiles and waited until second-generation missiles had been proven technologically before programming large forces of them. Strategic missile development led to a de-emphasis of manned strategic bombers, although the administration implemented some measures to improve bomber retaliatory capabilities immediately, such as alert programs, dispersal, and Positive Control. The triad—and the intellectual justification for it—remained a central element in U.S. strategic nuclear policy for decades.

Although the Eisenhower administration responded to strategic uncertainty by developing new types of delivery systems, it also approved high force levels for existing missile systems. Eisenhower programmed 810 ICBMs and 384 SLBMs during the missile gap period and allowed the level of bomber forces to decrease. While these missile force levels did not quite match the levels requested by the Air Force or the Navy, they exceeded the levels that, according to some estimates, would be necessary to impose unacceptable retaliatory damage on the Soviet Union in response to a Soviet first strike. In this period, Eisenhower approved the deployment of more strategic missiles than Kennedy or any subsequent president. Eisenhower's decisions meant that the major decision confronted by Kennedy was when to stop deploying missiles. Eisenhower and Kennedy established the force levels for strategic nuclear delivery vehicles for the next three decades.

President Eisenhower controlled his increase in the number and types of strategic forces by allowing only minor changes in the defense budget and in national strategy. He maintained stable defense budgets (despite pressure by the armed services, congressional Democrats, and the media) by making additional cuts in conventional forces. But extending the New Look formula was insufficient to keep defense budgets stable, so Eisenhower measured his strategic expansion and even cut programs for new bomber aircraft. His budgets fueled a public debate over whether the administration was allowing a "missile gap" to arise that would favor the Soviet Union.

President Eisenhower justified these budgetary decisions by maintaining a national strategy based on Massive Retaliation. In NSC debates concerning BNSP (NSCs 5810/1 and 5906/1) and nuclear war objectives (NSC 5904/1), Eisenhower steadfastly rejected any changes that might signal a decreased role for nuclear weapons and an increased one for conventional forces. He claimed that limited wars would inevitably escalate to a general (nuclear) war, so conventional forces would not be very useful. Eisenhower seemed to be determined to employ nuclear weapons if necessary, regardless of the locale. But there was a subtle evolution in Eisenhower's thinking with respect to massive retaliation. Gone were the coercive applications that nuclear superiority had allowed earlier in the decade. Eisenhower increasingly emphasized massive retaliation as a threat to commit mutual annihilation, not unlike mutual assured destruction. He never codified this transformation, however, largely because it could justify larger conventional forces and budgets. As a result, the New Look remained essentially unchanged throughout the Eisenhower presidency.

The continuation of massive retaliation as the basis of national strategy precipitated numerous difficulties for the Eisenhower administration. Many within and outside the government considered massive retaliation an incredible strategy. Eisenhower successfully fended off attacks on massive retaliation and on the resultant defense budgets, but he never reconstituted a public or governmental consensus in support of the strategy. Consequently, there was little debate when the Kennedy administration abandoned massive retaliation and implemented its policy of flexible response. The problems Eisenhower encountered in formulating nuclear strategy can also be attributed to the fact that massive retaliation provided no guidance as to the optimal strategic force posture or targeting doctrine. Thus, projections of expanding Soviet capabilities automatically justified, for the Air Force, longer target lists and larger strategic nuclear forces. As nuclear war plans grew to unimaginable levels of destruction, Eisenhower increasingly tried to control nuclear strategy as a way of limiting overkill and also limiting the demands on the defense budget from strategic forces. But his progress was hindered by interservice rivalry, which stymied even the most mundane JCS deliberations over nuclear strategy during the missile gap period—a rivalry that had arisen because of discontent with Eisenhower's budget priorities. Eisenhower's direct intervention in the formulation of nuclear strategy in 1960 came too late. Major decisions concerning nuclear strategy and targeting in an era of mutual vulnerability were thus left for the Kennedy administration.

Eisenhower's nuclear policy choices during the missile gap period constitute a somewhat mixed legacy. He transformed U.S. strategic forces when he ushered in the missile age and took important steps to reduce the forces' vulnerability, thus ensuring stable nuclear deterrence even after the Soviets deployed missiles. Eisenhower saved the nation billions of dollars by resisting pressure for higher defense spending. The concept of mutual assured destruction was also beginning to emerge, as evidenced in the administration's force posture decisions, the establishment of the SIOP, and the intellectual arguments developed by the WSEG and the PSAC. But Eisenhower's nuclear legacy from the missile gap period is also full of disappointments. Both national strategy and nuclear strategy were allowed to drift because of Eisenhower's retention of massive retaliation. His continued emphasis on nuclear forces rather than conventional forces, as well as his professed willingness to employ nuclear weapons for extended deterrence, increased the risk of a nuclear war fought over marginal interests.

This assessment of the nuclear policy choices made by Eisenhower during the missile gap period differs considerably from the one made popular by some Eisenhower revisionists.[2] Despite differences in focus and emphasis, these authors assert that Eisenhower's recognition of the horror, futility, and waste of nuclear war led him to fight a lonely but heroic rearguard action, after the Gaither report, to defrost the Cold War. These revisionists interpret Eisenhower's attempts to keep defense budgets stable as an indication of a greater desire to curtail the arms race. They often state this argument most explicitly when discussing Eisenhower's reaction to the Gaither report. Walter LaFeber writes: "But Eisenhower refused to panic. He flatly rejected most of Gaither's recommendations. . . . The retired general called military hardware this 'negative stuff.'"[3] Similarly, Stephen Ambrose comments that "he [Eisenhower] refused to bend to the pressure, refused to initiate a fallout shelter program, refused to expand conventional and nuclear forces, refused to panic. . . . It is doubtful if any other man could have done what Eisenhower did. The demands for shelters, for more bombers, for more bombs, for more research and development of missiles and satellites, was nearly irresistible. Only Ike could have gotten away with saying no. . . . But Eisenhower said no, and he kept saying no to the end of his term."[4]

In support of their argument, these revisionists stress Eisenhower's interest in arms control and diplomatic initiatives to improve superpower relations, especially during the missile gap period. Michael Beschloss claims that Eisenhower's "main aspiration was arms control. For seven years, he had hoped to move the world toward disarmament and had failed."[5] These authors attribute Eisenhower's failure to achieve arms control or improve diplomatic relations with the Soviet Union to such factors as bureaucratic resistance, the public's Cold War fears, Soviet misperception, and even simple bad luck. The revisionist perspective is evident in Ambrose's analysis of the collapse of the May 1960 summit, "If only Eisenhower had not given permission for that last flight. If only Khrushchev had not made such a big deal out of such a small thing. If only the two leaders could have trusted their own instincts just once, rather than their technicians and generals. . . . No one knows where the momentum [from a test ban agreement] thus generated might have taken the Cold War and the nuclear arms race."[6] Other authors emphasize a change in Eisenhower's thinking about nuclear war. Kenneth Thompson points out that Eisenhower "saw that thermonuclear destruction would be fundamentally unlike any larger-scale traditional war and undertook to check the unrealism of military and civilian leaders who spoke of 'prevailing' in a nuclear war."[7] Gregg Herken, citing PSAC member Jerome Wiesner, attributes this evolution in thinking to the influence of the Gaither report.[8]

Many Eisenhower revisionists reason that U-2 intelligence enabled Eisenhower to limit defense spending and make diplomatic initiatives without endangering national security. Because of the U-2, LaFeber writes, Eisenhower "knew no missile gap existed—or, more accurately, that one existed and that it overwhelmingly favored the United States."[9] Beschloss concludes that "for years, the President resisted unnecessary defense spending almost single-handedly against immense pressure. This was perhaps the principal achievement of the U-2 program and of his Presidency."[10] Thus, Eisenhower revisionists find that the President resisted further increases in defense spending and the arms race because of intelligence from the secret U-2 program.

This case finds that Eisenhower continued his New Look expansion of strategic forces within budgetary constraints because the U-2 intelligence was inconclusive and warning systems were precarious. Eisenhower rejected the force levels advocated by missile gap critics in Congress and the armed services, but his strategic force planning decisions indicate that in his view, reducing vulnerability had a higher priority than slowing the arms race. The U-2 provided unprecedented and reassuring intelligence, but its limitations prevented any conclusion that a Soviet missile gap would not exist within several years. The U-2 enabled Eisenhower and some top aides to question the predictions of NIEs, but there remained considerable uncertainty about the goals of Soviet missile programs. Had Eisenhower been more certain about Soviet missile strength, it is unlikely that he would have allowed worst-case scenarios to swirl throughout the administration or have authorized so many new missile programs.

Revisionists are correct to focus on Eisenhower's last three years in office because of the development of arms control thought within the government, the increased high-level superpower negotiations, and the movement toward a Nuclear Test Ban Treaty. There is no reason to question Eisenhower's desire for improved relations with the Soviet

Union and for a reduction in the arms race. But it would be a mistake to overemphasize early arms control initiatives and ignore the expansion of strategic nuclear forces and the continuation of the policy of massive retaliation. None of Eisenhower's arms control proposals would have affected U.S. nuclear superiority or foreclosed future avenues of weapons expansion.[11] For this reason, such measures had a limited appeal to the Soviets. As our case study shows, Eisenhower played an active and important role in expanding strategic nuclear forces. Even in his final year in office, Eisenhower approved adding hundreds of strategic missiles to the nation's strategic nuclear force, just as fears about the existence of a missile gap were beginning to subside.

The differences between my assessment of Eisenhower and the one offered by some Eisenhower revisionists might be explained in part by the many documents cited in this study that were declassified only in the past few years and thus were unavailable to many Eisenhower revisionists. Lacking access to documents that showed Eisenhower's commitment to continue the New Look and massive retaliation throughout the missile gap period, revisionists tended to focus on such issues as arms control and defense spending rather than on the formulation of strategic national strategy and force planning. They reconciled their new evidence concerning Eisenhower's attitudes on the arms race with the absence of results by arguing that he was undermined by forces beyond his control.

The Missile Gap Period and Its Lessons for the Presidency and Eisenhower Literatures

The literature reviewed in Chapter 1 represented competing schools of thought (Richard Neustadt and Graham Allison, post-Neustadt presidency scholars, Eisenhower contemporary or "conventional" school, and Eisenhower revisionists). While each school has shortcomings when applied to nuclear policy during the missile gap period, all are important for understanding this case.

This analysis of the making of U.S. nuclear policy during the missile gap period has concentrated on four policy areas: threat assessments, national and nuclear strategy, defense budgeting, and force planning. Although these areas are interrelated, they involve different processes and properties despite the obvious interrelationships. The unique properties of each policy area present different opportunities for, and obstacles to, presidential control. Many scholars have noted the differences between policy areas (see Chapter 1), but their typologies do not seem sufficiently sensitive in this case.[12] Presidential power varies across policy areas due to three variables. The first is the constitutional distribution of power in the policy area. Policies in which the president and Congress share power increase the requirements and opportunities for bargaining.[13] Generally, Congress has deferred to the president's constitutional authority in foreign policy—as it did in matters of national security policy during the Cold War. In this analysis, we can distinguish between policy areas by whether President Eisenhower's decisions required the formal approval of Congress.[14] The second variable is the complexity of the issue area. Issues that are very complex may encourage the president's dependence on others for special expertise, technical information, or analysis in order to make policy in that area. The institutional location of such information may also be important. An overdependence on such information by presidents can threaten their policy-making power in numerous ways, including receiving biased information, the inability to receive the information desired, or simply obstructionism to delay the policy process. The third variable is the degree to which presidents depend on others to implement their policies faithfully. The nature of the decision and the type of behavior desired can either aggravate or mitigate implementation problems. Graham Allison argues that policy decisions entail action, thus inviting coordination problems and implementation that is inconsistent with decision makers' intents.[15] Robert Art asserts that not all policies suffer from "slippage" during implementation, and presidential oversight can limit it.[16] Decisions that do not require action, or certain types of actions, pose few implementation problems.

These three variables enable us to detect the continuum of different policy areas. At one extreme would be policies in which decisions are not subject to congressional approval; require little technical information, expertise, or analysis; and are not likely to encounter implementation problems. Such a policy offers the best opportunity for the unfettered exercise of presidential power, resembling Richard Neustadt's "self-executing" decisions or simply presidential command power.[17] At the other extreme would be policies in which a presidential decision requires congressional approval; forces decision makers to rely on others for expertise, information, and analysis; and presents extensive implementation difficulties. Policy making in these areas obviously entails many obstacles to the exercise of presidential power. Successful presidential leadership under these conditions requires the most skillful manipulation of the bargaining advantages accorded the office. Between these two extremes are, of course, policy areas that possess these properties in varying degrees. Policies in this category may also demand presidential leadership that differs from the other two types.

Of the policies examined in this study, national strategy is the one which is closest to presidential command power. Eisenhower's BNSP decisions were made without congressional approval, depended minimally on specialized information,[18] and presented no immediate implementation requirements.[19] Consequently, national strategy making was concentrated in the NSC and was dominated by Eisenhower. While nuclear strategy was closely related to national strategy, it displays very different policy characteristics. Nuclear strategy was not subjected to congressional examination. In attempting to formulate nuclear strategy, however, Eisenhower depended on specialized analyses of nuclear warfare by the armed services. The military's resistance to production of this information obstructed and delayed nuclear strategy decision making throughout 1958 and 1959. Ensuring that the military, particularly the Air Force, revised nuclear war plans according to presidential decisions was an enormous oversight task—one that the Eisenhower administration was not entirely successful in performing. The implemention of nuclear strategy decisions continued to be such a problem that Eisenhower took special measures in 1960 to monitor the development of the first SIOP. Thus, nuclear strategy is a classic example of presidential power in foreign policy making: freedom from congressional restraints but dependence on the bureaucracy for information critical for making decisions, and for faithful implementation.

Defense budgeting and strategic nuclear force planning are similar with respect to the three variables. Decisions in both policy areas usually required congressional approval, but a few of Eisenhower's force planning decisions, such as the impoundment of funds, did not. Congressional review meant an expanded debate and more opportunity for bargaining, but the advantages Eisenhower enjoyed in these areas protected his policy choices. Eisenhower depended to a great extent on specialized technical information and analysis in the making of these policy decisions. Defense budget information and weapons analysis for use in force planning were generated by the military and scientific communities and were the basis of administration policy deliberations in these areas. Neither of these policy areas presented major problems in implementation, since there was little room for "slippage." Once Congress had appropriated the funds, the armed services could not spend more or reallocate those funds for other purposes (although the administration could and did). Similarly, the services could not develop new weapons systems or increase force deployments without the administration's approval.

Threat assessment, based largely on intelligence estimates, involves yet another set of policy dynamics. This area is highly complex and requires the sophisticated analysis of very technical information. President Eisenhower lacked the staff and expertise to conduct anything other than a cursory evaluation of intelligence estimates. Thus Eisenhower, and to a lesser extent members of Congress, could either believe such reports or ignore them, but there was no immediate decision to be made and no implementation issue. Threat assessment based on intelligence estimates demonstrates the extent to which the bureaucracy can generate technical information for policy makers in a manner that allows them little choice but to accept them. Even if presidents attach little credence to them, intelligence reports are still circulated throughout the administration, shaping actors' perspectives and framing the discussions of still other reports and policies.[20] Because of the importance of these reports and the absence of decision, presidents must exercise great care in their decisions that relate to structuring how information is formulated and distributed, as well as its function in the policy process.

The obstacles to the successful exercise of presidential power are not equally distributed across policy areas as the closely related areas of defense budgeting and nuclear strategy testify. Neustadt and Allison emphasized these obstacles and the extent to which presidents must depend on others if they are to be successful in policy making. But because both authors assumed that presidents want policy to result in action, they overlooked the possibility that the structural features of some policy areas might accentuate presidential power rather than limit it.

If we turn to the policy-making process, we find that nuclear strategy during the missile gap period was influenced by organizational factors and bargaining, consistent with the perspectives of Allison and Neustadt. National strategy formulation in 1958 and 1959 resembles

Allison's governmental politics model in several respects. The positions that were taken by many—although by no means all—participants seem to have been determined by organizational interests, confirming that "where you stand is where you sit." The coalition-building process that frequently develops among subordinates and continues up the bureaucratic ladder was evident in the BNSP debates, particularly in 1958. Organizational interests and bureaucratic conflict also influenced administration deliberations about defense budgeting and strategic force planning. However, Allison's third model is insufficient for understanding this case. Most significantly, policy outcomes do not appear to be "resultants" of a bargaining process but rather were a conscious choice made by President Eisenhower.

The limited value of Allison's governmental politics model can be attributed to Eisenhower's leadership skills. Eisenhower contained bureaucratic bargaining and protected his decision-making authority by utilizing White House institutions and his managerial practices. He employed both informal meetings with key advisers and formal NSC meetings to advance his policy goals.[21] His frequent informal Oval Office meetings allowed Eisenhower to keep track of issues, mull over ideas and alternatives, vent his anger, and hammer out decisions. The informal process was not limited to the making of national security policy but was an integral part of Eisenhower's policymaking. He rarely made policy decisions in formal meetings such as the weekly NSC meetings, and when he did it was usually on issues of secondary importance—language in a policy paper, for example. The formal processes of the NSC, particularly the weekly meetings, served Eisenhower's policy purposes in other ways. He instilled collective responsibility among top administration officials by bringing policy decisions to the NSC and ensuring their inclusion in the record of NSC actions. Eisenhower sometimes used the meetings to remind his administration of what he would not tolerate—in effect, putting officials on notice. The NSC meetings also gave officials the opportunity to communicate their particular concerns about policies to the president. While this almost never led to changes in policies, it effectively allowed officials to vent disagreements and helped to unite the administration behind the president's decisions. The value for Eisenhower of using weekly NSC meetings to remind administration officials of his constitutional authority should not be underestimated.

Eisenhower's skill in using White House institutions to advance his policies is illustrated in the President's Science Advisory Committee and the Office of the Special Assistant for Science and Technology. Created shortly after Sputnik, they gave Eisenhower access to the highest quality scientific advice free of organizational bias. These scientists, especially the special assistants (initially, James Killian, who was replaced by George Kistiakowsky), played an increasingly important role throughout the missile gap period as Eisenhower drew on their analyses in the making of policy, especially with regard to defense budgeting and strategic force planning. Eisenhower relied on the PSAC for information in his attempt to challenge the military. Thus, as military issues became more scientifically complex, Eisenhower created a new organization independent of the military that he could trust.[22]

Eisenhower buttressed his use of White House organizations with three managerial practices to strengthen his hand further in administration policymaking. First, he influenced policy through his ability to determine who participated in deliberations—manipulating what Allison calls the "action-channels." The informal advisory system proved especially effective in this regard. Officials not included in Eisenhower's informal meetings would not be privy to his perspective, and thus would be at a disadvantage as policies moved forward. Participation in NSC discussions and those its of subordinate boards could hardly compensate for not having access to Eisenhower and the highest-level deliberations. Defense budgeting provides a striking example of how Eisenhower used his control of action-channels to his advantage. In each year, he excluded the armed services from administration deliberations after they had made their initial recommendations for the fiscal year. This reduced bargaining while the administration was finishing its budget submission. The services were brought back into the process only after the draft was completed; they were, in effect, presented with a fait accompli.

Second, Eisenhower controlled policy making by establishing the parameters of debates at early stages. His unparalleled understanding of U.S. military policy gave him the knowledge and confidence necessary to intervene early in policy making and to communicate his objectives. In both the formulation of national strategy and defense budgeting, the president shaped policy outcomes by telling aides what changes he would accept. This framed policy deliberations and curtailed political bargaining, since administration officials did not seriously consider options that they knew Eisenhower would not endorse.[23]

Third, Eisenhower used his appointment power to ensure that administration officials did his bidding and not that of the organization in which they served. Although administration appointees in defense areas tended to be nonpolitical experts, none of these men were inclined to question the President's judgment. Eisenhower was particularly well served by his White House staff, which dutifully provided him with the information and analysis necessary for making decisions. Staffers such as Robert Cutler, Gordon Gray, James Killian, George Kistiakowsky, and Andrew Goodpaster had no agenda except to serve the president. In the formulation of national strategy, the White House staff assisted Eisenhower without limiting his decision-making authority. Every year, Cutler and Gray sought revision of the BNSP only to have the idea rejected by the president. They accepted this verdict, recognizing that such rejection was the president's prerogative and that their role was only to bring policy options to his attention. The behavior of the White House staff during the Eisenhower administration is refreshing.

This assessment of Eisenhower's leadership style and use of White House institutions conforms closely with the analyses of a number of presidency scholars and Eisenhower revisionists. Some presidency scholars, particularly John Burke and Terry Moe, have argued that White House institutions are essential for presidential success.[24] Without such supporting mechanisms, presidents would have a difficult time overcoming obstacles to their goals inside and outside the government. Eisenhower created a White House system that promoted his objectives, protected his authority to make decisions, monitored the policy-making process, and provided him with independent expertise; in this way he kept his administration united and focused. It was a system that fit Eisenhower perfectly and made the most of his knowledge of military affairs. It is doubtful that this arrangement would be as effective for another president. Not all aspects of Fred Greenstein's "hidden-hand" style are evident in this study due in part to the nature of these issues. But Eisenhower's reliance on formal and informal structures, emphasis on maintaining the unity of his administration, and desire to portray himself as above politics while being political are aspects of Greenstein's argument which are illustrated in this case. Still other management and leadership practices available to all presidents, such as making appointments and controlling access, were used by Eisenhower to strengthen his position in administration policy making. This supports assertions by Robert Art and Stephen Krasner that

Allison overlooked the president's opportunities to limit bureaucratic bargaining.[25]

This assessment of the effectiveness of Eisenhower's leadership style within the administration with regard to the formulation of nuclear policy should not be interpreted as a rejection of contemporary criticism by the Eisenhower "conventional" school. In this respect, this case differs from the analyses of most Eisenhower revisionists. Eisenhower's success inside the administration was not matched outside the administration. In the one area that required congressional approval, the defense budget, Eisenhower's authority in military affairs and congressional deference prevented any major changes in his program. Congress, the opinion elites, and the general public expected changes in defense strategy and weapons, but these materialized either slowly or not at all. Thrust into a national debate over the missile gap and defense policy, Eisenhower and administration officials failed to convince these groups of the validity of their policies. This point was made in the analyses of a number of scholars of the conventional school. Samuel Huntington wrote that "the President has failed to take the initiative in bringing strategic issues to the people, in arousing support for foreign and military proposals, and in educating the public to its responsibilities in the nuclear age."[26] Norman Graebner and Richard Neustadt, among others, worried that the nation might be ill-prepared for problems that might arise in the areas of both foreign and domestic policy because of Eisenhower.[27]

Contemporary accounts remind us of the liabilities of Eisenhower's leadership style. Presidential success in the public debate over defense policy during the missile gap period demanded strong public and political leadership, not apolitical, hidden-hand maneuvering. But this was a role Eisenhower detested and denigrated. There was no place he could turn for help, since his White House was structured to reinforce his hidden-hand preferences. Consequently, Eisenhower struggled when forced into a realm of presidential leadership that was unfamiliar and uncomfortable. In this respect, Neustadt's criticism of Eisenhower's apolitical temperament is entirely justified. Neustadt's observations about the relationship between presidential power, public opinion, and professional reputation are confirmed.[28] Even though Eisenhower was successful in dealing with the bureaucracy and Congress on narrow issues of nuclear policy, his inability to sway the public in the debate over the missile gap undermined his political power in the long run. It was a failure of political leadership attributable to the leadership style that worked so well within the administration.

There is no one approach in the literatures on the presidency and Eisenhower that is adequate for understanding Eisenhower's successes and failures in the making of nuclear policy during the missile gap period. Presidential power in administration policy making is constrained by organizational factors and bureaucratic bargaining. Their effect varies depending on the policy because of the complexity of issues and also because of implementation requirements. Problems of implementation, particularly "slippage," can be limited by whether the president's decision requires action and by the nature of the action. A president has some tools at his disposal to overcome or minimize these limitations, including White House organization, the power of appointments, and other managerial practices. These tools will be most valuable if they accentuate the president's personal strengths and compensate for his weaknesses. But such tools are less helpful to a president in persuading the Congress and public to support—either explicitly or implicitly—his policies. Here a president must rely on his political skills and political capital to carry the day. Whereas policy making within the administration demands that presidents be executives, success in policy making outside the administration requires them to be politicians. The balance between the two roles shifts depending on the policy area; most of the important events in the making of nuclear policy took place within the administration and required Eisenhower to be an executive. Presidents may gravitate toward the role that is more comfortable or more consistent with their own perception of the office. But because presidents are not in complete control of their agendas, they must be prepared to play both roles. Ineffectiveness in playing either role may lead to policy setbacks in the short term and, depending on the number and gravity of the failures, may undermine their presidency in the long term.

Notes

1. The Eisenhower Administration, Presidential Power, and Nuclear Policy

1. Security Resources Panel of ODM-SAC, *Deterrence and Survival in the Nuclear Age* (hereafter referred to as the "Gaither report"), U.S. Cong., Joint Committee on Defense Production, 94th Cong., 2d sess. (Washington, D.C.: GPO, 1976), p. 16.

2. The report stated that "the main protection of our civil population against a Soviet nuclear attack has been and will continue to be the deterrent power of our armed forces, to whose strengthening and securing we have accorded the highest relative value. But this is not sufficient unless it is coupled with measures to reduce the extreme vulnerability of our people and our cities. As long as the U.S. population is wide open to Soviet attack, both the Russians and our allies may believe that we shall feel increasing reluctance to employ SAC in any circumstances other than when the United States is directly attacked." Ibid., pp. 18, 34.

3. The missile gap period begins with the Sputnik I launch and lasts until 1961, when U.S. intelligence finally concluded that the Soviet Union would not be able to develop and deploy missiles quickly enough to gain a numerical advantage over the United States.

4. The 1990 edition of *Presidential Power* contains the eight original chapters and the five chapters added in subsequent editions. Richard E. Neustadt, *Presidential Power and the Modern Presidents* (New York: Free Press, 1990).

5. Ibid., p. 24. Neustadt described "self-executing" commands as having five components: "On each occasion the President's involvement was unambiguous. So were his words. His order was widely publicized. The men who received it had control of everything needed to carry it out. And they had no apparent doubt of his authority to issue it to them." Ibid., p. 18.

6. Ibid., p. 28.

7. "The essence of a President's persuasive task is to convince such men that what the White House wants of them is what they ought to do for their sake and on their authority." Ibid., p. 30.

8. Ibid., p. 71.

9. "The President-as-teacher has a hard and risky job. If he wants to guard his popular approval he must give real-life experience a meaning that will foster tolerance for him. But happenings create his opportunities to teach. He has to ride events to gain attention." Ibid., p. 89.

10. Ibid., p. 150.

11. The president "mounts guard, as best he can, when he appraises the effects of present action on the sources of his influence. In making that appraisal he has no one to depend on but himself; his power and its sources are a sphere of expertise reserved to him." Ibid., p. 152.

12. Ibid., p. 11.

13. In 1966, Neustadt and Allison were members of the same Harvard University study group that investigated the bureaucracy's effect on policy making. Graham T. Allison, *Essence of Decision: Explaining the Cuban Missile Crisis* (Boston: Little, Brown, 1971), pp. ix, 147–149.

14. The major works by these authors that Allison cited were: Samuel P. Huntington, *The Common Defense* (New York: Columbia University Press, 1961); Huntington, "Strategy and the Political Process," *Foreign Affairs* 38 (January 1960):285–299; Warner Schilling, Paul Hammond, and Glenn Snyder, *Strategy, Politics, and Defense Budgets* (New York: Columbia University Press, 1962); and Roger Hilsman, "The Foreign-Policy Consensus: An Interim Report," *The Journal of Conflict Resolution* 3 (December 1959):361–382.

15. Allison, *Essence of Decision*, p. 144.

16. Ibid., p. 176.

17. "The rules of the game stem from the Constitution, statutes, court interpretations, executive orders, conventions, and even culture. Some rules are explicit, others implicit. Some rules are quite clear, others fuzzy. Some are very stable; others are ever changing. But the collection of rules, in effect, defines the game." Ibid., p. 170.

18. Ibid., p. 175.

19. Robert Art divides scholars of foreign policy making in two groups. The "first wave" includes Neustadt, Schilling, Hilsman, and Huntington, whereas the "second wave" includes Allison, and Morton Halperin. His article is still the finest discussion of the differences between these two groups. Robert J. Art, "Bureaucratic Politics and American Foreign Policy: A Critique," *Policy Sciences* 4 (1973):467–490.

20. Allison, *Essence of Decision*, p. 162.

21. See Terry Moe, "Presidents, Institutions, and Theory," in George C. Edwards III, John H. Kessel, and Bert A. Rockman, eds., *Researching the Presidency: Vital Questions, New Approaches* (Pittsburgh, Pa.: University of Pittsburgh Press, 1993), pp. 337–386.

22. Art, "Bureaucratic Politics and American Foreign Policy," p. 476; Peter Sperlich, "Bargaining and Overload: An Essay on *Presidential Power*," in Aaron Wildavsky, ed., *Perspectives on the Presidency* (Boston: Little, Brown, 1975), p. 419; Stephen Krasner, "Are Bureaucracies Important? (Or Allison Wonderland)," *Foreign Policy* 7 (1971): 166. See also David Welch, "The Organizational Process and Bureaucratic Politics Paradigms: Retrospect and Prospect," *International Security* 17 (1992):112–146.

23. See John P. Burke, *The Institutional Presidency* (Baltimore: Johns Hopkins University Press, 1992), especially chap. 3.

24. Richard Tanner Johnson, *Managing the White House: An Intimate Study of the Presidency* (New York: Harper & Row, 1974); Alexander L. George, *Presidential*

Decisionmaking in Foreign Policy: The Effective Use of Information and Advice (Boulder, Colo.: Westview, 1980).

25. "Indeed one might plausibly imagine a president prone to formalistic modes of decision making deliberately selecting a more collegial pattern (or vice versa) to offset his own personal weaknesses and those of the system he might otherwise favor, even at the price of stress and strain. . . . But there does seem to be evidence that presidents whose staff systems closely reflect their personal predilections fare less well than presidents who are open to change or modification." Burke, *The Institutional Presidency*, p. 192.

26. Ibid., pp. 179–180.

27. "At precisely the same time that Neustadt's work was reorienting scholarly thinking around the concept of the personal presidency, the presidency itself was becoming highly institutionalized. Indeed, the hallmark of the modern presidency is its growth and development as an institution." Moe, "Presidents, Institutions, and Theory," p. 340. He observes that "the institutional presidency . . . is how presidents fight back against a system that is stacked against them." Ibid., p. 372.

28. John p. Burke, "The Institutional Presidency," in Michael Nelson, ed., *The Presidency and the Political System*, 3d ed. (Washington, D.C.: Congressional Quarterly Press, 1990), pp. 383–408.

29. In *The Institutional Presidency* (p. 194), Burke writes: "But while well-crafted staff structures and procedures and astute informal staff relations are likely to contribute to the president's thoughtful and informed choices, they cannot guarantee wise policy choices."

30. This article and the subsequent debate about its thesis are included in Steven A. Shull, ed., *The Two Presidencies: A Quarter Century Assessment* (Chicago: Nelson-Hall, 1991).

31. Lowi distinguished between distributive, redistributive, and regulatory policies. Theodore J. Lowi, "American Business, Public Policy, Case Studies, and Political Theory," *World Politics* 16 (1964):677–715.

32. Paul C. Light, "Presidential Policy Making," in Edwards, Kessel, and Rockman, *Researching the Presidency*, pp. 161–199. See also Light, *The President's Agenda: Domestic Policy Choice from Kennedy to Reagan (with Notes on George Bush)*, rev. ed. (Baltimore: Johns Hopkins University Press, 1991).

33. See, for example, Hugh Heclo, "Issue Networks and the Executive Establishment," in Anthony King, ed., *The New American Political System* (Washington, D.C.: American Enterprise Institute, 1978), pp. 87–124.

34. Scholarly perspectives on Eisenhower and nuclear weapons are discussed in Chapter 6.

35. Fred I. Greenstein, "Dwight D. Eisenhower: Leadership Theorist in the White House," in Greenstein, ed. *Leadership in the Modern Presidency* (Cambridge: Harvard University Press, 1988), p. 76.

36. Neustadt, *Presidential Power*, p. 138.

37. Ibid., p. 139.

38. Ibid., pp. 68–69.

39. Eisenhower "became typically the last man in his office to know tangible details and the last to come to grips with acts of choice. . . . Within its [Eisenhower's staff system] limits it was reasonably well-designed and rather more than reasonably effective. In Eisenhower's instance there is no disputing that it may have been essential to his health and peace of mind. But its workings often were disastrous for his hold on personal power." Ibid., p. 133.

40. Ibid., p. 152.

41. The success of the "amateur" Eisenhower clearly frustrated Neustadt: "The striking thing about our national elections in the fifties was not Eisenhower's personal popularity; *it was the genuine approval of his candidacy by informed Americans whom one might have supposed would know better*" (emphasis added). Ibid., pp. 162–163

42. Sidney Hyman, "The Eisenhower Glow Is Fading Away," *Reporter*, September 19, 1957, p. 15.

43. Norman Graebner, "Eisenhower's Popular Leadership," in Dean Albertson, ed., *Eisenhower as President* (New York: Hill and Wang, 1963), pp. 147–159. Graebner also yearned for an activist president: "What matters far more in Presidential success are, first, the intellectual alertness necessary to penetrate contemporary movements and, second, the political craftsmanship required to translate victory into political action which meets the challenge of the times. Measured by its adaptation of the Democratic past to the conditions of the 1950's, the Eisenhower leadership had been a success, indeed an historical necessity. But the permanent judgment of that leadership will hinge on the President's achievement in influencing, within the limits of his power, the fundamental trends of this age toward the protection of this nation's well-being." Ibid., p. 159.

44. Richard H. Rovere, *Affairs of State: The Eisenhower Years* (New York: Farrar, Straus, and Cudahy, 1956), p. 353.

45. Ibid., p. 355.

46. "There is a great deal of which Eisenhower has never heard, and he has organized his office staff and his Cabinet into a kind of conspiracy to perpetuate his unawareness. In no administration within memory have so few decisions been made by the President himself. . . . The President has made it known that he wants to become part of the administrative process only when those below him have reached unanimity. . . . Eisenhower has not, of course, wished to cripple his own imagination and sensitivity, but the mechanical apparatus is of his own making and the insulation deliberately sought." Ibid., pp. 356–357.

47. Ibid., pp. 370–371. To Rovere, Eisenhower offered "the spectacle, novel in the history of the Presidency, of a man strenuously in motion yet doing essentially nothing—traveling all the time yet going nowhere." Quoted in Graebner, "Eisenhower's Popular Leadership," p. 155.

48. Murray Kempton, "The Underestimation of Dwight D. Eisenhower," *Esquire*, September 1967, p. 108; and Garry Wills, *Nixon Agonistes: The Crisis of the Self-Made Man* (Boston: Houghton Mifflin, 1969).

49. Referring to the Taiwan Straits crisis, Rovere wrote: "A stronger-willed President could have avoided altogether some of the crises that plagued the Eisenhower administration." *Affairs of State*, p. 365.

50. Ibid., p. 364.

51. Ibid., p. 367.

52. Merlo J. Pusey, *Eisenhower the President* (New York: Macmillan, 1956), pp. 285–286, 294–295.

53. Samuel Lubell, *Revolt of the Moderates* (New York: Harper & Brothers, 1956), pp. 24, 25, 26, 28.

54. According to Lubell (p. 30), "One aide who has watched this 'softening up' technique employed repeatedly attributes the tactic to the fact that 'Eisenhower likes people and tries not to hurt anyone.' But others in the White House say

quite bluntly, 'The President is as skilled a maneuverer as the Army has produced.'"

55. Ibid., pp. 30, 32–34, 35, 37.

56. This is the title of Lubell's chapter on Eisenhower. Lubell recognized that this leadership style might prevent Eisenhower from making politically unpopular but necessary policiy decisions (pp. 49–50).

57. For surveys of this literature, see Vincent DeSantis, "Eisenhower Revisionism," *Review of Politics* 38 (1976):190–207; Mary McAuliffe, "Commentary: Eisenhower, the President," *Journal of American History* 68 (1981):625–631; Anthony James Joes, "Eisenhower Revisionism and American Politics," in Joann Krieg, ed., *Dwight D. Eisenhower: Soldier, President, Statesman* (Westport, Conn.: Greenwood Press, 1987), pp. 283–296; Richard H. Immerman, "Confessions of an Eisenhower Revisionist: An Agonizing Reappraisal," *Diplomatic History* 14 (1990):319–342; and Stephen Rabe, "Eisenhower Revisionism: A Decade of Scholarship," *Diplomatic History* 17 (1993): 97–115.

58. Immerman, "Confessions of an Eisenhower Revisionist," p. 319.

59. Fred I. Greenstein, *The Hidden-Hand Presidency: Eisenhower as Leader* (New York: Basic Books, 1982), p. 57. For the discussion of the six strategies, see pp. 58–92. Other works on Eisenhower's management style are: Greenstein, "Eisenhower as an Activist President: A New Look at the Evidence," *Political Science Quarterly* 94 (Winter 1979–1980):575–599; Greenstein, "Dwight D. Eisenhower"; John W. Sloan, "The Management and Decision-Making Style of President Eisenhower," *Presidential Studies Quarterly* 20 (1990):295–313; Sloan, *Eisenhower and the Management of Prosperity* (Lawrence: University Press of Kansas, 1991); Burke, *The Institutional Presidency*, chap. 3; Burke and Greenstein, *How Presidents Test Reality* (New York: Russell Sage Foundation, 1989); and Kenneth Thompson, "The Strengths and Weaknesses of Eisenhower's Leadership," in Richard Melanson and David Mayers, eds., *Reevaluating Eisenhower: American Foreign Policy in the Fifties* (Chicago: University of Illinois Press, 1987), pp. 13–30.

60. Greenstein, *The Hidden-Hand Presidency*, chap. 4; Sloan, "The Management and Decision-Making Style of President Eisenhower."

61. The similarities between Lubell and these Eisenhower revisionists are remarkable. I have yet to find a revisionist who refers to Lubell's work, however. This is not entirely surprising because most revisionists rely on only a handful of works in describing contemporary evaluations of Eisenhower.

62. For example, see Gordon H. Chang, "To the Nuclear Brink," *International Security* 12 (1988):96–122.

63. See John Lewis Gaddis, *Strategies of Containment: A Critical Appraisal of Postwar American National Security Policy* (New York: Oxford University Press, 1982); Richard Betts, *Nuclear Blackmail and Nuclear Balance* (Washington, D.C.: Brookings Institution, 1987); and Roger Dingman, "Atomic Diplomacy during the Korean War," *International Security* 13 (1988–1989):50–91.

64. David Alan Rosenberg, "The Origins of Overkill: Nuclear Weapons and American Strategy, 1945–1960," *International Security* 7 (1983):3–71.

65. Ibid., pp. 12–27. The most thorough discussion of command and control is Peter Douglas Feaver, *Guarding the Guardians* (Ithaca: Cornell University Press, 1992).

66. Harry Borowski, *A Hollow Threat: Containment and Strategic Air Power before Korea* (Westport, Conn.: Greenwood Press, 1982); and Peter Roman, "Curtis LeMay

and the Origins of NATO Atomic Targeting," *Journal of Strategic Studies* 16 (1993):46–74.

67. Marc Trachtenberg, "A 'Wasting Asset': American Strategy and the Shifting Nuclear Balance, 1949–1954," *International Security* 13 (1988–1989):11–18.

68. Gaddis, *Strategies of Containment*, pp. 127–129. For an excellent description of the 1952 presidential campaign, see Stephen E. Ambrose, *Eisenhower*, vol. 1: *Soldier, General of the Army, President-Elect, 1890–1952* (New York: Simon & Schuster, 1983), chaps. 26 and 27.

69. Stephen E. Ambrose, *Eisenhower*, vol. 2: *The President* (New York: Simon & Schuster, 1984), pp. 20–24.

70. Burke, *The Institutional Presidency*, p. 61.

71. Ibid., pp. 60–61; and Greenstein, *The Hidden-Hand Presidency*, pp. 127–128.

72. See Robert Cutler to Eisenhower, memorandum, March 16, 1953, "Subject: Recommendations Regarding the National Security Council," in Department of State, *Foreign Relations of the United States* (hereafter cited as *FRUS*), *1952–1954*, vol. 2: *National Security Affairs* (Washington, D.C.: GPO, 1984), p. 245. Eisenhower had three special assistants for national security affairs during his eight years as president: Robert Cutler (1953–1955, 1957–1958), Dillon Anderson (1955–1956), and Gordon Gray (1958–1961).

73. See Ambrose, *Eisenhower*, 2:217.

74. Sloan, "The Management and Decision-Making Style of President Eisenhower," pp. 302–304; and Greenstein, *The Hidden-Hand Presidency*, p. 134.

75. Gaddis, *Strategies of Containment*, pp. 146–147; McGeorge Bundy, *Danger and Survival* (New York: Random House, 1988), pp. 246–247; Samuel F. Wells, Jr., "The Origins of Massive Retaliation," *Political Science Quarterly* 96 (1981):44.

76. Cutler, Memorandum for Record, May 9, 1953, "Subject: Solarium Project," in *FRUS, 1952–1954*, 2:323–326. Richard Immerman reports that the administration thought about having a fourth group study preventive war but dropped the idea. Immerman, "Confessions of an Eisenhower Revisionist," p. 337. See also Robert Amory to Allen Dulles, memorandum, July 8, 1953, "Subject: Project 'Solarium'," *Declassified Documents Catalog*, 1989, document 3157 (hereafter cited as *DDC* with year/document number).

77. For documents relating to Project Solarium, see *FRUS, 1952–1954*, 2:378–440. Solarium is also discussed in H.W. Brands, "The Age of Vulnerability: Eisenhower and the National Insecurity State," *American Historical Review* 94 (1989):966–968.

78. Gaddis, *Strategies of Containment*, pp. 147–148.

79. National Security Council, "Basic National Security Policy," NSC 162/2, October 29, 1953, in *FRUS, 1952–1954*, 2:578–597. Quotations are from paragraphs 11, 34, and 39.

80. A debate within the government followed NSC 162/2 about the proper interpretation of the paper, particularly the sections on custody and operational planning. Eisenhower eventually decided that the military should have more custody of weapons and should be allowed to plan to employ them. According to the NSC secretary "The paragraph [39–b] is not a decision in advance that atomic weapons will in fact be used in the event of *any* hostilities. In certain cases the use of nuclear weapons by the United States would be automatic. . . . Many situations, however, involve political questions of the gravest importance which cannot be precisely foreseen." James Lay to Secretary of State et al., memorandum, January 4, 1954,

"Subject: Policy Regarding Use of Nuclear Weapons," *DDC*, 1988/2266. See also Bundy, *Danger and Survival*, pp. 249–250.

81. Eisenhower mentioned the doctrine several days earlier in his State of the Union message. See Wells, "The Origins of Massive Retaliation," p. 33.

82. John Foster Dulles, "The Evolution of Foreign Policy," in Philip Bobbitt, Lawrence Freedman, and Gregory F. Treverton, eds., *U.S. Nuclear Strategy* (New York: New York University Press, 1989), p. 124.

83. See Bernard Brodie, *Strategy in the Missile Age* (Princeton, N.J.: Princeton University Press, 1959), pp. 248–263; William W. Kaufmann, "The Requirements of Deterrence," in Kaufmann, ed., *Military Policy and National Security* (Princeton, N.J.: Princeton University Press, 1956), pp. 12–38; Anthony Buzzard, "Massive Retaliation and Graduated Deterrence," *World Politics* 8 (1956):228–37; Robert Endicott Osgood, *Limited War* (Chicago: University of Chicago Press, 1957).

84. National Security Council, "Basic National Security Policy," NSC 5602/1, March 15, 1956, in *FRUS, 1955–1957*, vol. 19: *National Security Policy* (Washington, D.C.: GPO, 1990), pp. 242–257, esp. paras. 11–15.

85. Thomas B. Cochran, William M. Arkin, Robert S. Norris, and Jeffrey I. Sands, *The Nuclear Weapons Databook*, vol. 4: *Soviet Nuclear Weapons* (New York: Ballinger, 1989), pp. 25, 40.

86. *Alert Operations and the Strategic Air Command, 1957–1991* (Offutt Air Force Base, Nebraska: SAC Office of the Historian, 1991), p. 79.

87. See Edmund Beard, *Developing the ICBM* (New York: Columbia University Press, 1976); Michael Brown, *Flying Blind: The Politics of the U.S. Strategic Bomber Program* (Ithaca: Cornell University Press, 1992); and Harvey Sapolsky, *The Polaris System Development* (Cambridge: Harvard University Press, 1972).

88. See Feaver, *Guarding the Guardians*, chap. 7.

89. There were three situations in which a commander in chief could declare a Defense Emergency: A "major" communist attack on U.S. or Allied forces; an attack on the continental United States; and the declaration of an "Air Defense Readiness" or "Air Defense Emergency." AEC-DOD, "Memorandum of Understanding for the Transfer of Atomic Weapons," May 4, 1956, *DDC*, 1992/71.

90. The report assumed an average yield of 80 kilotons per bomb. National Security Council, "Summary Evaluation of the Net Capability of the U.S.S.R. to Inflict Direct Injury on the United States Up to July 1, 1955," NSC 140/1, May 18, 1953, in *FRUS, 1952–1954*, 2:328–334.

91. National Security Council, "Tentative Guidelines under NSC 162/2 for FY 1956," NSC 5422, June 14, 1954, in *FRUS, 1952–1954*, 2:651.

92. "Soviet Capabilities and Probable Soviet Courses of Action through 1960," NIE 11-3-55, May 17, 1955, *Declassified Documents Quarterly Catalog* (hereafter cited as *DDQC*), 1978/22B.

93. See Robert Bowie to John Foster Dulles, memorandum, June 13, 1956, "NSC Consideration of Policy on Continental Defense (NSC 5606)"; and memorandum, "Subject: Discussion at the 288th Meeting of the NSC, June 15, 1956." In *FRUS, 1955–1957*, 19:316–320.

94. Joint Intelligence Committee to Secretaries, note, January 27, 1957, "Soviet Capabilities and Probable Courses of Action against North America in a Major War Commencing during the Period 1 January 1958 to 31 December 1958," JIC 491/122, *DDQC*, 1981/169A.

95. See Betts, *Nuclear Blackmail and Nuclear Balance*, pp. 147–148.

96. NSC 140/1, in *FRUS, 1952–1954*, 2:335.

97. NSC 5422; NSC 5422/2 "Guidelines under NSC 162/2 for FY 1956," August 7, 1954; and James Lay to the NSC, memorandum, October 11, 1954, "Subject: Summary Statement of Existing Basic National Security Policy." In *FRUS, 1952–1954,* 2:651, 725, 755.

98. See "Editorial Note," in *FRUS, 1955–1957,* 19:78.

99. NIE 11-3-55, *DDQC,* 1978/22B.

100. John Prados, *The Soviet Estimate* (New York: Dial Press, 1982), chap. 4.

101. "World Situation and Trends," NIE 100-5-77, November 1, 1955; and Memorandum, "Discussion at the 268th Meeting of the NSC on December 1, 1955." In *FRUS, 1955–1957,* 19: 134, 169.

102. Department of State, Memorandum, "Subject: U.S. and Soviet Missiles," ibid., pp. 154–161.

103. Prados, *The Soviet Estimate,* p. 46.

104. Betts, *Nuclear Blackmail and Nuclear Balance,* p. 158.

105. "Estimate of the Situation," annex to NSC 5602/1, in *FRUS, 1955–1957,* 19:259.

106. NSC 140/1, in *FRUS,* 1952–1954, 2:348.

107. Eisenhower, "Diary Entry by the President, January 23, 1956"; report of the Panel on the Human Effects of Nuclear Weapons Development; November 21, 1956; and memorandum, "Subject: Discussion at the 306th Meeting of the NSC, December 20, 1956." In *FRUS, 1955–1957,* 19:187, 374, 380.

108. Kenneth Shaffel, *The Emerging Shield: Air Defense, 1945–1960,* (Washington, D.C.: Office of Air Force History, 1991).

109. Memorandum, "Subject: Discussion at the 306th Meeting of the NSC, December 20, 1956," in *FRUS, 1955–1957,* 19:380.

110. See report of the Technical Capabilities Panel to Eisenhower, February 14, 1955, "Meeting the Threat of Surprise Attack," ibid., pp. 41–56; and Project RAND, *Protecting U.S. Power to Strike Back in the 1950's and 1960's,* Report R-290 (Santa Monica: Rand Corporation, 1956).

111. NESC 1956 claimed that "with the advent of long-range ballistic missiles, the present concept of the military and civil defense of the United States against nuclear weapons will require extensive revision in view of the drastically reduced amount of warning time available against 'Surprise Attack.'" Memorandum, "Subject: Discussion at the 306th Meeting of the NSC, December 20, 1956," p. 380.

112. "Implications of Growing Nuclear Capabilities for the Communist Bloc and the Free World," NIE 100-5-55, June 14, 1955, in *FRUS, 1955–1957,* 19: 86.

113. Quoted in memorandum, "Subject: Discussion at the 314th Meeting of the NSC, February 28, 1957," ibid., p. 427.

114. See Roman, "Curtis LeMay"; Robert Wampler, *NATO Strategic Planning and Nuclear Weapons, 1950–1957,* Nuclear History Program Occasional Paper no. 6 (College Park: University of Maryland School of Public Affairs, Center for International Security Studies, 1990); and Marc Trachtenberg, *History and Strategy* (Princeton, N.J.: Princeton University Press, 1991).

115. Jacob Neufeld, *The Development of Ballistic Missiles in the United States Air Force, 1945–1960* (Washington, D.C.: Office of Air Force History, 1990), pp. 222–227.

116. See Feaver, *Guarding the Guardians,* chap. 8.

117. For insight into the debate over NATO at the end of the Eisenhower administration see the report by Robert Bowie. For the declassified report, see Robert Bowie, *The North Atlantic Nations' Tasks for the 1960's,* Nuclear History Program

Occasional Paper no. 7 (College Park: University of Maryland School of Public Affairs, Center for International Security Studies, 1991).

118. See Betts, *Nuclear Blackmail and Nuclear Balance*, pp. 76–79; Trachtenberg, *History and Strategy*, chap. 5; and Ambrose, *Eisenhower*, vol. 2: *The President*, pp. 517–518.

2. Soviet Missiles, Warning, and Perceptions of Vulnerability

1. Andrew J. Goodpaster, Memorandum of Conference with the President (MCP) [on October 8, 1957, at 8:30 A.M.], October 9, 1957, Missiles and Satellites 1(3) (September–December, 1957) folder, Subject series, DOD subseries, White House Office, Office of the Staff Secretary (WHO-OSS) Records, Dwight D. Eisenhower Library (DDEL).

2. Robert Cutler, Notes of Conference in the President's Office [October 8, 1957, at 8:30 A.M.], Missiles and Satellites 1(3) (September–December, 1957) folder, Subject series, DOD subseries, WHO-OSS, Records, DDEL.

3. See CIA Consultants to Allen Dulles, letter, October 23, 1957, and enclosure to letter, Allen Dulles to Goodpaster, October 28, 1957. In *DDQC*, 1980/8B. See also Allen Dulles and Herbert Scoville, "Briefing for Preparedness Investigating Subcommittee of the Armed Services Committee of the Senate on Soviet Guided Missiles and Related Soviet Capabilities," November 26–27, 1957, *DDQC*, 1984/1544, p. 29.

4. Memorandum, "Subject: Discussion at the 339th Meeting of the NSC, October 10, 1957," October 11, 1957, 339th Meeting of the NSC folder, NSC series, Ann C. Whitman (ACW) file, DDEL.

5. CIA Consultants to Allen Dulles, letter, October 23, 1957, enclosure to letter, Allen Dulles to Goodpaster, October 28, 1957, *DDQC*, 1980/8B. The identity of the consultants is sanitized in the original document.

6. Security Resources Panel of ODM-SAC, *Deterrence and Survival in the Nuclear Age* (Gaither report), U.S. Congress, Joint Committee on Defense Production, 94th Cong. 2d sess. (Washington, D.C.: GPO, 1976), pp. 26–27.

7. Security Resources Panel of ODM-SAC, "Advisors' Notes for Conference, 4 November 1957," *DDQC*, 1984/523, p. 3.

8. "Main Trends in Soviet Capabilities and Policies 1957–1962," NIE 11-4-57, November 12, 1957, *DDQC*, 1979/128A, p. 27.

9. Bruce D. Berkowitz and Allan E. Goodman, *Strategic Intelligence for American National Security* (Princeton, N.J.: Princeton University Press, 1989), p. 123.

10. Allen Dulles and Herbert Scoville, "Briefing for Preparedness Investigating Subcommittee," pp. 29, 30, 42.

11. National Security Council, "U.S. Policy on Continental Defense," NSC 5802/1, February 19, 1958, *DDQC*, 1979/414A, p. 2.

12. John Prados, *The Soviet Estimate* (New York: Dial Press, 1982), p. 82; and Allen Dulles to Chairman, Guided Missile Intelligence Committee, memorandum, October 9, 1958, "Subject: Soviet ICBM Development Program," *DDQC*, 1984/1545.

13. Allen Dulles to Chairman, Guided Missile Intelligence Committee, October 9, 1958.

14. The earliest possible deployments could have been in late 1958 but only if the

Soviets put the program on a "crash" basis. Goodpaster, Memorandum for Record, June 9, 1958, June 1958 folder, Nuclear History series, National Security Archive, Washington, D.C. See also Prados, *The Soviet Estimate*, p. 78.

15. Senator Stuart Symington to Eisenhower, letter, August 29, 1958, *DDQC*, 1978/118B.

16. One of Lanphier's sources claimed that since the beginning of 1957, the Soviet Union had flight-tested 55 long-range missiles, but a second source asserted that the number could be as high as 80. These sources also believed that the Soviets had begun construction of operational launch facilities at Murmansk and Kamchatka, and had even tested nuclear warheads during missile flight tests. Ibid.

17. Howard Stoertz, Jr., Memorandum for the Record, August 18, 1958, "Subject: Discussion of Soviet and U.S. Long-Range Ballistic Missile Programs," *DDQC*, 1981/421A.

18. Symington to Eisenhower, letter, August 29, 1958.

19. Referring to the still very secret U-2 program, President Eisenhower told Symington that "Colonel Lanphier cannot possibly get information that reflects the full process of the evaluation system." Goodpaster, MCP [on August 29, 1958], August 30, 1958, *DDQC*, 1981/641B.

20. Dulles also pointed out: "We believe that the Soviet ballistic missile program has emphasized reliability and simplicity, rather than extreme refinement of design." Allen Dulles to Eisenhower, memorandum, October 10, 1958, "Subject: Evaluation of Information on Soviet Ballistic Missile Capabilities, contained in a letter to the President from Senator Stuart Symington dated 29 August 1958," *DDQC*, 1984/1546.

21. Specifically: "Is our intelligence coverage sufficient to establish with a high degree of confidence the number of Soviet ICBM and earth satellite firings, including the number of failures? . . . What is the likelihood that the apparent slow rate of ICBM test firing represents serious difficulty and delay in the development program? . . . In sum, what is the most likely explanation for the observed pattern of Soviet ICBM development to date, and what modifications, if any, should be made to our existing estimate?" Allen Dulles to Chairman, Guided Missile Intelligence Committee, memorandum, October 9, 1958.

22. Also attending the briefing were Assistant Air Force Chief of Staff for Intelligence Major General James Walsh, Guided Missile Intelligence Committee Chairman Colonel Earl McFarland, Jr. (USAF), Herbert Scoville, and Howard Stoertz, Jr. Lanphier attended only part of the meeting because he had no governmental affiliation.

23. Allen Dulles, Memorandum of Conversation, December 16, 1958, "Subject: DCI Briefing of Senator Symington on Soviet Ballistic Missile Programs and Capabilities," *DDQC*, 1982/2350.

24. Ibid., p. 5.

25. The administration had concluded in June that Soviet bomber deployments had fallen far short of CIA predictions. See Goodpaster, Memorandum for Record, June 9, 1958, and Goodpaster, MCP [on June 17, 1958], June 18, 1958, both documents located in June 1958 folder, Nuclear History series, National Security Archive.

26. Allen Dulles, Memorandum of Conversation, December 16, 1958, pp. 5–6.

27. Symington and Lanphier tried to distance themselves from their earlier accusations on several occasions during the meeting. Still, the two tenaciously argued their case and demanded access to the CIA's raw intelligence data. The senator

claimed that he "assumed the numbers of firings he had reported were extreme, and that he had quoted them to get the DCI to take a look at the intelligence system." Ibid., pp. 5, 8.

28. Ibid., p. 9.

29. Lawrence McQuade to Paul Nitze, memorandum, May 31, 1963, "Subject: But Where Did the Missile Gap Go?" enclosure to memorandum, Nitze to McGeorge Bundy, May 30, 1963, Missile Gap (February–May 1963) folder, Subject series, National Security File (NSF), John F. Kennedy Library (JFKL).

30. Ibid.

31. Weapons Systems Evaluation Group, "The Feasibility, Cost, and Effectiveness of Dual Runways at SAC Bomber Bases," Staff Study no. 77 (WSEG-SS-77), July 20, 1959, *DDQC*, 1981/161.

32. See Allen Dulles to Eisenhower, memorandum, October 10, 1958.

33. CIA Deputy Director for Intelligence Robert Amory explained that "it isn't a major operation, but it is such a marginal weapon that we doubt they would make that decision." Testimony of Allen Dulles before the Senate Foreign Relations Committee, January 26, 1959, U.S. Congress, Senate Committee on Foreign Relations, *Executive Sessions of the Senate Foreign Relations Committee* (historical series), vol. 12, 86th Cong., 1st sess. (Washington, D.C.: GPO, 1982), p. 104.

34. General Nathan F. Twining, Presentation before the Senate Armed Services Committee, January 20, 1959, p. 7, supplied in response to author's Freedom of Information Act request.

35. These would constitute some of the first Soviet deployments of nuclear-propelled submarines, according to the CIA. Allen Dulles reported: "We have had reports from secret intelligence sources indicating that they have 12 nuclear-powered submarines either finished or under construction, but we have no confirmation of it." Dulles testimony on January 26, 1959 before the Senate Foreign Relations Committee, p. 97.

36. Prados, *The Soviet Estimate*, p. 84; and Lawrence Freedman, *U.S. Intelligence and the Soviet Strategic Threat*, 2d ed. (Princeton, N.J.: Princeton University Press, 1986), p. 70.

37. "There is no evidence to support or deny the present availability of ten IOC missiles." L. A. Hyland to Director of Central Intelligence Allen Dulles, memorandum, August 25, 1959, "Subject: Report of DCI Ad Hoc Panel on Status of the Soviet ICBM Program," *DDQC*, 1984/776.

38. The members of the Panel worried that this absence might be due to inadequate intelligence collection. Ibid.

39. Chairman of the Joint Chiefs of Staff Nathan Twining to Defense Secretary Neil McElroy, memorandum, November 30, 1959, CCS 471.94 folder, CJCS Records, Record Group (RG) 218, Modern Military Branch (MMB), National Archives.

40. General Nathan Twining, Presentation before the Subcommittee on the Department of Defense, House Committee on Appropriations, January 13, 1960, *DDC*, 1989/2476, p. 5.

41. CJCS Twining to Defense Secretary McElroy, memorandum, November 30, 1959.

42. CJCS Twining told Defense Secretary McElroy: "There have been four positive sightings of these submarines. . . . All sightings have been in silhouette; hence, no firm evidence of possible missile tubes or tube covers is available." Ibid.

43. See testimony of Allen Dulles before the Seuate Foreign Relations Committee, January 18, 1960, U.S. Cong., Senate Committee on Foreign Relations, *Executive*

Sessions of the Senate Foreign Relations Committee (historical series), vol. 12 (1960), 86th Cong. 2d sess. (Washington, D.C.: GPO, 1982), p. 8.

44. See McQuade to Nitze, memorandum, May 31, 1963; and report, "The Feasibility and National Security Implications of a Monitored Agreement to Stop or Limit Ballistic Missile Testing and/or Production," March 4, 1960, *DDQC*, 1984/533, p. 22. See also memorandum, "Subject: Discussion at the 433rd Meeting of the NSC on January 21, 1960," *DDC*, 1991/1949.

45. Testimony of Allen Dulles before the Senate Armed Services Committee, January 18, 1960, Senate Committee, *Executive Sessions*, p. 10; General Nathan Twining Presentation before the Subcommittee on the Department of Defense of the House Committee on Appropriations, January 13, 1960, p. 3.

46. See report, "The Feasibility and National Security Implications," p. 22.

47. McQuade to Nitze, memorandum.

48. George B. Kistiakowsky, *A Scientist at the White House*, (Cambridge, Harvard University Press, 1976), p. 219.

49. Report, "The Feasibility and National Security Implications," p. 22.

50. CJCS Twining presentation, January 13, 1960. See also Allen Dulles testimony before Senate Foreign, Relations Committee, January 18, 1960, Senate Committee, *Executive Sessions*, p. 11.

51. Ibid.

52. Raymond L. Garthoff, *Assessing the Adversary* (Washington, D.C.: Brookings Institution, 1991), p. 41 no. 109.

53. Weapons Systems Evaluation Groups, "Initial Study of Arms Control Measures Affecting the Risk of Surprise Attack," WSEG Report no. 52 (hereafter cited as WSEG-52), January 6, 1961, p. 61, supplied in response to author's Freedom of Information Act (FOIA) request.

54. McQuade to Nitze, memorandum, May 31, 1963.

55. Dino A. Brugioni, *Eyeball to Eyeball* (New York: Random House, 1991), p. 53.

56. Testimony of Allen W. Dulles to Senate Foreign Relations Committee, May 31, 1960, Senate Committee, *Executive Sessions*, p. 284.

57. Freedman, *U.S. Intelligence and the Soviet Strategic Threat*, p. 69. In *Eyeball to Eyeball* (p. 43), Brugioni lists seven targets of Powers's flight.

58. This limitation is evident in the following table:

Percentage of time suitable for aerial photography, on the average, USSR, by region (percentages)

Region	Winter	Spring	Summer	Fall
European U.S.S.R.	5–10%	5–10%	5–30%	5–10%
Siberia:				
Western	25–40	15–25		5–10
Central	15–25	30–50 (SC)	5 (SE) 10 (SC) 15 (NC)	20–40

Source: WSEG Report no. 52, p. 188.
SE = Southeast; SC = South Central; NC = North Central.

59. Dulles testimony, on May 31, 1960, Senate Committee, *Executive Sessions*, p. 308.

60. Garthoff, *Assessing the Adversary*, p. 42.

61. Allen Dulles to Eisenhower, letter, October 10, 1958.

62. By 1959, the standard interpretation of Soviet ICBM objectives was: "the indications are that the Soviet ICBM program is not a crash program designed to provide a considerably greater number than we have estimated. But we believe it is designed to provide a substantial ICBM capability at an early date." Testimony of Allen Dulles before Senate Foreign Relations Committee, January 18, 1960, Senate Foreign Relations Committee, *Executive Sessions*, p. 13.

63. Berkowitz and Goodman, *Strategic Intelligence*, pp. 128–131.

64. Thomas B. Cochran, William M. Arkin, Robert S. Norris, and Jeffrey I. Sands, *The Nuclear Weapons Databook*, vol. 4: *Soviet Nuclear Weapons* (New York: Ballinger, 1989), p. 106.

65. Thomas Wolfe, *Soviet Power in Western Europe* (Baltimore: Johns Hopkins University Press, 1970), p. 41.

66. See "Active Defense against Ballistic Missiles," Part D, vol. 1, enclosure to the Gaither report, Security Resources Panel vol. 1(5) folder, NSC series, Subject subseries, WHO-OSANSA, DDEL.

67. NSC 5802/1; see also recommendations, President's Board of Consultants on Foreign Intelligence Activities to Eisenhower, October 24, 1957, *DDC*, 1992/489.

68. Jerome Wiesner and H. J. Watters to George Kistiakowsky, memorandum, August 10, 1959, "Subject: Report of the Results of the 29–30 July 1959 Meeting of the Ad Hoc Panel on Continental Air Defense," Air Defense (August 1958–December 1960) (1) folder, White House Office—Office of Special Assistant for Science and Technology (WHO-OSST,) DDEL.

69. Lieutenant General J. H. Atkinson, commander of the Air Defense Command, to commander in chief, Strategic Air Command (CINCSAC), memorandum, October 24, 1957, "Subject: SAC Alert Warning Times"; appendix to memorandum, chief of staff, United States Air Force (CSUSAF) to the JCS, January 16, 1958, "SAC Alert Warning Times," enclosure to note, Secretaries to JCS, JCS 1899/385, January 21, 1958, *DDQC*, 1981/155A.

70. Kenneth Schaffel, *The Emerging Shield* (Washington, D.C.: Office of Air Force History, 1991), p. 223.

71. Wiesner and Watters to Kistiakowsky, memorandum, August 10, 1959.

72. The Gaither report's study of warning, stated: "In the immediate future the only choice seems to be a considerable expansion of our present activities—even though it is not clear what this will buy. . . . In summary, there is no reason whatsoever to feel complacent about the effectiveness of our present warning system. Once again the problem seems to be 'too little and too late.'" "Early Warning," enclosure to the Gaither report, section I, part B, vol. 1, pp. B-14, B-15, Security Resources Panel, in vol. 1 (2) folder, NSC series, Subject subseries, WHO-OSANSA, DDEL.

73. This proposal was very similar to a 1955 plan. Schaffel, *The Emerging Shield*, p. 255.

74. William Minshull, Memorandum for Record, May 21, 1958, "Subject: Status of BMEWS," in AICBM, Early Warning (November 1957–December 1960) folder, WHO-OSST, DDEL.

75. WSEG-SS-77, p. 22.

76. Acting Deputy Chief of Staff/Operations, United States Air Force to Air Force Chief of Staff, memorandum, January 15, 1960, "Subject: BMEWS," in Missile/Space/Nuclear folder, Thomas White Papers, Library of Congress.

77. "There also exist known methods that can greatly decrease the effectiveness

of BMEWS radars even against reasonably large numbers of missiles." "Discussion Paper on 'Current Issues in the Philosophy of Continental Defense,'" November 11, 1960, in Air Defense (August 1958–December 1960) (2) folder, WHO-OSST, DDEL. On September 27, 1961, twelve days after the first BMEWS began operating, CINCSAC Thomas Power wrote: "I am concerned greatly over the prospects of having to rely upon a system which has serious deficiencies." Power to LeMay, cable, September 27, 1961, in Messages and Cables (September 1961) folder, Curtis LeMay Papers, Library of Congress.

78. NSC 5802/1. The Gaither study on early warning concluded: "our present and planned capabilities for detecting and identifying submarines approaching our coasts are frighteningly unreliable and thoroughly inadequate." "Early Warning," enclosure to the Gaither report, p. B-13.

79. Because SOSUS is still in operation, most details concerning it are classified. A few details have emerged as a result of budget constraints since the end of the Cold War. See William Broad, "Scientists Oppose Navy Plan to Shut Undersea Monitor," *New York Times*, June 12, 1994, p. 1.

80. "Presentation by the Director, Weapons Systems Evaluation Group, to the National Security Council on the Subject of Offensive and Defensive Weapons Systems," October 13, 1958, supplied in response to author's FOIA request. The Air Force remained suspicious of the capabilities of SOSUS as late as 1963. An Air Force deputy chief of staff estimated that SOSUS "has a probability of only 0.15 to 0.25 of detecting a snorkeling or noisy submarine within 1,000 nautical miles off the U.S. coast for each detection opportunity. Its capability against quiet submarines is practically nil at long ranges (missile launch ranges)." Lieutenant General Disosway to LeMay, memorandum, February 28, 1963, "Detection of Submerged Submarines," in 1963 Air Staff Actions folder, LeMay Papers, Library of Congress.

81. David Beckler to James Killian, memorandum, May 8, 1959, "Subject: Aide-Mémoire on Major Actions of the President's Science Advisory Committee, November 1957–May 1959," *DDQC*, 1977/205A.

82. Power to White, letter, December 23, 1957; and Power to White, letter, November 4, 1957. In 1957 Top Secret, General file, White Papers, Library of Congress.

83. CJCS Twining to McElroy, memorandum, March 28, 1958, "Subject: Antisubmarine Effort," *DDQC*, 1983/851.

84. Twining to Gates, memorandum, November 20, 1959, "Subject: Early Warning against Submarine-Launched Ballistic Missiles," in CJCS 471.94 folder, CJCS Records, RG 218, MMB, National Archives.

85. John C. Toomay, "Warning and Assessment Sensors," in Ashton B. Carter, John D. Steinbruner, and Charles A. Zraket, eds., *Managing Nuclear Operations* (Washington, D.C.: Brookings Institution, 1987), p. 297.

86. "U.S. Policy on Continental Defense," NSC 5802/1. February 19, 1958.

87. Gregg Herken, *Counsels of War* (New York: Alfred A. Knopf, 1985), pp. 123–125.

88. "For example, NIC [National Indicators Center] is in daily touch with Soviet troop movements in East Germany, but these reach Washington about 48 hours after such movements are observed." Jerome Wiesner to James Killian, memorandum, May 12, in 1959, in AICBM-Early Warning (November 1957–December 1960) folder, WHO-OSST, DDEL.

89. Atkinson to Power, memorandum, October 24, 1957.

90. Power to White, letter, October 21, 1957, in October 1957 folder, Nuclear History series, National Security Archive.

91. Robert Cutler to Eisenhower, handwritten notes on memorandum, October 25, 1957, "Subject: SAC Concentration in U.S. and Reaction Time (as of above date)," in Security Resources Panel (2) (1957–1958) folder, NSC series, Briefing Notes subseries, WHO-OSANSA, DDEL.

92. Brigadier General T. R. Milton to LeMay, memorandum, October 31, 1957, in Air Force (3) folder, Subject series, Confidential file, DDEL.

93. Gaither report.

94. Goodpaster, MCP [on November 7, 1957], November 7, 1957, *DDQC*, 1979/331A.

95. "I said I considered that such an attack without provocation involving casualties of perhaps one hundred million would be so abhorrent to all who survived in any part of the world that I did not think that even the Soviet rulers would dare to accept the consequences." John Foster Dulles, MCP [on November 7, 1957], November 7, 1957, *DDQC*, 1984/1630. On January 3, 1958, Robert Sprague met with John Foster Dulles. Sprague told him: "During the next 2-1/2 years (more or less) the U.S. position vis-a-vis the Soviet Union will be at its strongest. During this period we can knock out the Soviet Union's military capability without taking a similar blow from the Soviet Union. Our present capability to do this is increasing. During this period the Soviet Union could in retaliation hurt the U.S., but not put us out of action." John Foster Dulles, Memorandum of Conversation, January 3, 1958, *DDQC*, 1984/1631.

96. Weapons Systems Evaluation Group, "First Annual Review of WSEG Report no. 23—The Relative Military Advantages of Missiles and Manned Aircraft," August 8, 1958, pp. 10–11, in Targeting and "The Relative Military Advantages of Missiles and Manned Aircraft" folder, NSC series, Briefing Notes subseries, WHO-OSANSA, DDEL.

97. "Presentation by the Director, Weapons Systems Evaluation Group, to the National Security Council," October 13, 1958.

98. Brockway McMillan, "An Analysis of Technical Factors in the Strategic Posture of the United States — 1956–64," March 4, 1959, in Air Defense (October 1958–August 1959) folder, WHO-OSST, DDEL.

99. Ibid.

100. Goodpaster, MCP [on March 4, 1959, 2:30 P.M.], March 19, 1959, *DDQC*, 1977/352D.

101. The WSEG chose 1962 as the target year because that would be the first year in which dual runways, BMEWS, and an SLBM warning system would be available to the United States. By 1962, SAC would have 62 bases in the zone of interior to support 630 B-52s, 990 B-47s, 90 B-58s, and more than 1,100 tanker aircraft. The WSEG assumed that there would be ground or runway alert for only one-third of SAC and no airborne alert. Further, the WSEG believed that the size of the alert force could be increased from one-third to one-half if the United States received a warning time of 12 hours instead of 15 minutes. Report WSEG-SS-77, pp. 2–3, 10.

102. "If there is no such [warning] system in existence during the time period of this study, it would be possible for the enemy to attack SAC with no tactical warning whatsoever and the expected number of aircraft launched for perfectly executed attacks would be a function only of the number of randomly surviving bases, possibly amounting to less than 20 percent of the total alert force." Ibid., pp. 21, 38.

103. Ibid., p. 39.

104. Although Eisenhower said "he did not believe that when the Soviets got all

their missiles ready, they would turn them loose against us," he admitted that "we ought to assume that the Soviets will make the first attack." "Memorandum, Subject: Discussion at the 430th Meeting of the NSC, January 7, 1960," *DDC*, 1991/3345.

105. Herbert F. York, *Making Weapons, Talking Peace* (New York: Basic Books, 1987), p. 188.

106. Kistiakowsky, *A Scientist at the White House*, p. 262.

107. "Until technology permits the deployment of an effective active defense against submarine-launched ballistic missiles, the principal measures of protection should be provided by the capability to attack prior to launch." National Security Council, "U.S. Policy on Continental Defense," NSC 6022/1, cited in Robert Johnson to McGeorge Bundy, memorandum, January 30, 1961, "Subject: The FY 1962 Budget—Issues Relating to the U.S. Military Program," in Department of Defense, FY 1963 (January–October 1961) folder, Departments and Agencies series, National Security file, JFKL.

108. Weapons Systems Evaluation Group, "Evaluation of Strategic Offensive Weapons Systems," Report no. 50, December 27, 1960, p. 7, in December 1960 folder, Nuclear History series, National Security Archive.

109. Quoted in Johnson to Bundy, memorandum, January 30, 1961.

110. See attachment to memorandum, H. Sidney Buford III to Gordon Gray, November 21, 1960, "Subject: Trends in the Power Positions of the U.S. and the U.S.S.R. and of the Free World and the Sino-Soviet Bloc, 1960–1970," *DDC*, 1987/2366.

111. WSEG Report no. 52, pp. 63, 65, 66.

3. The Challenge to Massive Retaliation

1. National Security Council, "Basic National Security Policy," NSC 5707/8, June 3, 1957, in *FRUS, 1955–1957*, vol. 19: *National Security Policy* (Washington, D.C.: GPO, 1990), p. 511.

2. National Security Council, "Basic National Security Policy," NSC 162/2, October 30, 1953, in *FRUS, 1952–1954*, vol. 2: *National Security Policy* (Washington, D.C.: GPO, 1984), p. 593.

3. Martin Navias, *Nuclear Weapons and British Strategic Planning, 1955–1958* (New York: Oxford University Press, 1991).

4. The other two points, reaction time of nuclear forces and whether strategic systems should be designed to disarm the Soviet Union or only retaliate, are discussed in the next section.

5. Attachment to letter, William Foster to Killian, January 13, 1958, *DDQC*, 1981/511A.

6. When officials of the Atomic Energy Commission and the State and Defense Departments approached Eisenhower in August 1957 to obtain his approval for a test of a reduced-radiation or "clean" nuclear warhead, Cutler argued that the new weapons might shore up declining credibility. He gave Eisenhower two critical evaluations of U.S. policy. A JCS staffer stated that "our faith in them [large nuclear weapons] should not be so overriding as to blind us to our needs for other instruments of policy, both state and military." Captain Jack Morse, AEC representative to the NSC, claimed, "I cannot believe we would invoke world holocaust for our allies; if we say we will our allies will not believe us, and Russia will soon prove the threat an idle one." Enclosures to memorandum, Cutler to Eisenhower, August 7, 1957,

"Subject: Limited War in the Nuclear Age," in Cutler, 1956–1957 (1) folder, Administration series, ACW file, DDEL.

7. Joint Secretariat to the JCS, note, December 30, 1957, "The Character and Probable Results of a Future General War," JCS 2280/2, *DDQC*, 1981/153A.

8. See F. M. Dearborn, Jr., to Eisenhower, memorandum, February 14, 1958, in Nuclear Policy (1958) folder, NSC series, Briefing Notes subseries, WHO-OSANSA, DDEL. USIA Director Arthur Larson alerted AEC Chairman Lewis Strauss to the propaganda value of clean weapons shortly after the launch of Sputnik. See Larson to Strauss, letters, October 15, 1957, *DDC*, 1990/2099, 2100.

9. Naval studies of the problem of credibility aroused Cutler's interest and probably were the source of some of these arguments. Cutler to Christian Herter, letter, March 17, 1958, *DDQC*, 1987/1177. On Jack Morse, see S. T. Cohen, *The Neutron Bomb: Political, Technological, and Military Issues* (Cambridge: Institute for Foreign Policy Analysis, 1978), pp. 4–16.

10. Morse to Cutler, memorandum, "Subject: Massive Deterrent," March 8, 1958, enclosure to letter, Morse to Cutler, March 22, 1958, in Nuclear Policy (1958) folder, NSC series, Briefing Notes subseries, WHO-OSANSA, DDEL.

11. Cutler to Hoegh, memorandum, March 21, 1958, *DDQC*, 1988/2369.

12. Memorandum, "Subject: Discussion at the 359th Meeting of the NSC, Thursday, March 20, 1958," March 21, 1958, *DDC*, 1990/310.

13. Cutler, memorandum, "'Clean' Nuclear Weapons," March 16, 1958, *DDQC*, 1982/1542 and 1981/632A. These are the same document but with different sections sanitized. Goodpaster, Memorandum for Record, March 21, 1958, in March 1958 folder, Nuclear History series, National Security Archive; Cutler, notes, March 21, 1958, *DDQC*, 1981/632A.

14. Hans Bethe to Killian, memorandum, April 17, 1958, *DDC*, 1991/2231.

15. Memorandum, "Subject: Discussion at the 360th Meeting of the NSC, Thursday, March 27, 1958," March 28, 1958, *DDC*, 1990/311.

16. John Lewis Gaddis argues that Dulles' views approximated a flexible response approach. See John Lewis Gaddis, "The Unexpected John Foster Dulles: Nuclear Weapons, Communism, and the Russians," in Richard H. Immerman, ed., *John Foster Dulles and the Diplomacy of the Cold War* (Princeton, N.J.: Princeton University Press, 1990), p. 54.

17 John Foster Dulles, MCP [on April 1 1958], April 1, 1958, *DDC*, 1989/3430.

18. Goodpaster, Memorandum for record, April 9, 1958, *DDQC*, 1982/1577.

19. Enclosure to letter, Cutler to John Foster Dulles, April 7, 1958, *DDC*, 1987/3587.

20. Morse continued to influence Cutler's ideas on the these issues. He wrote Cutler: "So long as we offer only the danger of world destruction to defend Europe, so long will the whole world doubt our sincerity or our sanity. . . . In Europe [the president] might encourage NATO to stand in its own defense, using our nuclear weapons if necessary, and without reliance upon American strategic nuclear forces to strike the Soviet homeland." Morse to Cutler, memorandum, April 9, 1958, "Subject: Why American Leadership of the Free World Declines," in Review of BNSP, 1954–1960 (2) folder, NSC series, Briefing Notes subseries, WHO-OSANSA, DDEL.

21. The Planning Board based its review on the annual "Estimate of the World Situation" (NIE 100–58), as well as the advice of five "outside" consultants: John McCloy, General Alfred Gruenther, Karl Bendetson, Robert Bowie, and Arthur Burns.

22. National Security Council, "Basic National Security Policy," NSC 5810, April 15, 1958, *DDQC*, 1980/286B.

23. Cutler to John Foster Dulles, letter, May 2, 1958, in State Department (2) (January 1958–July 1959) folder, Special Assistant series, Subject subseries, WHO-OSANSA, DDEL.

24. Cutler, "Alternative Version of Paragraph 14," May 1, 1958, *DDQC*, 1980/378B.

25. W. J. McNeil to McElroy, memorandum, April 25, 1958, "Subject: Comments on NSC 5810, 'BNSP'," in Review of BNSP, 1954–1960 (2) folder, NSC series, Briefing Notes subseries, WHO-OSANSA, DDEL.

26. Twining to McElroy, memorandum, April 25, 1958, "Subject: BNSP—NSC 5810," in CCS 381 folder, CJCS Records, RG 218, NA.

27. Cutler, "Major Factors Influencing Review of Basic Policy," May 1, 1958, *DDQC*, 1980/378A.

28. Subsequent discussion of this NSC meeting is from this source as well. Memorandum, "Subject: Discussion at the 364th Meeting of the NSC, Thursday, May 1, 1958," May 2, 1958, in 364th NSC Meeting on May 1, 1958, folder, NSC series, ACW file, DDEL.

29. Maxwell Taylor later wrote that Dulles had failed to provide the support he had expected at this meeting. It is unclear how Taylor could have made such an assertion, given the statements by Dulles that follow. In any event, Dulles was dead and the NSC transcript remained classified "top secret" when Taylor made this claim. See Maxwell Taylor, *The Uncertain Trumpet* (New York: Harper & Brothers, 1960), pp. 59–65.

30. Cutler to McElroy, letter, May 5, 1958, *DDC*, 1990/2377.

31. White to Lt. General John K. Gerhart, memorandum, May 29, 1958, in May 1958 folder, Nuclear History series, National Security Archive.

32. John Foster Dulles to Eisenhower, cable, May 7, 1958, in Dulles May 1958 folder, Dulles-Herter series, ACW file, DDEL.

33. Cutler, notes, "Conference—June 17, 1958," *DDQC*, 1982/1578.

34. McElroy to the NSC, memorandum, June 18, 1958, "Subject: Basic National Security Policy," *DDQC*, 1980/357A.

35. Memorandum, "Subject: Discussion at the 370th Meeting of the NSC, June 26, 1958," June 27, 1958, *DDC*, 1990/356.

36. John Foster Dulles to Eisenhower, letter, July 23, 1958, *DDC*, 1988/2219.

37. Upon Eisenhower's approval, NSC 5810 became NSC 5810/1. Memorandum "Subject: Discussion at the 373rd Meeting of the NSC, Thursday, July 24, 1958," July 25, 1958, *DDC*, 1990/330.

38. Goodpaster, MCP [on July 14, 1959], July 15, 1959, in DOD 3(7) (July–August 1959) folder, Subject series, DOD subseries, WHO-OSS, DDEL.

39. "The danger to U.S. security from the Communist threat lies not only in general war or local aggression but in the possibility of a future shift in the East-West balance of power. Such a shift could be caused by a gradual erosion of Western positions via means short of force, and over time by a continued growth of over-all Communist strength at a rate significantly greater than that of the West." National Security Council, "Basic National Security Policy," NSC 5906, June 8, 1959, pp. 3–5, *DDC*, 1986/1577,

40. Gray, MMP [on Monday, May 18, 1959, at 9:00 A.M.], May 19, 1959, *DDC*, 1986/1072.

41. Annex to NSC 5906, NSC 5906, p. 61.

42. Gray, MMP [on Wednesday, July 15, 1959, at 11:35 A.M.], July 17, 1959, in Meetings with the President, June–December, 1959 (5) folder, Special Assistant series, Presidential subseries, WHO-OSANSA, DDEL.

43. Gray, MMP [on Monday, July 6, 1959, at 11:10 A.M.], July 8, 1959, in Meetings with the President, June–December, 1959 (5) folder, Special Assistant series, Presidential subseries, WHO-OSANSA, DDEL.

44. Memorandum, "Subject: Discussion at the 415th Meeting of the NSC, Thursday, July 30, 1959," July 30, 1959, *DDC*, 1990/999.

45. Memorandum, "Subject: Discussion at the 412th Meeting of the NSC," July 9, 1959, *DDC*, 1990/987.

46. National Security Council, "Basic National Security Policy," NSC 5906/1, August 5, 1959, *DDQC*, 1983/1305.

47. Reminding the angry president of the importance of this BNSP paper, Gray said that "it would be unlikely that he would wish to change this language next year and that a new administration would likely not rush in with major changes in basic policy at the beginning and that therefore we were writing language for the next two or three years, or possibly more. Therefore, it [was] necessary that the language be clear and definitive and understood and accepted by all." Gray, MMP [on Wednesday, July 15, 1959, at 11:35 A.M.], July 17, 1959, *DDQC*, 1985/2142.

48. At one NSC meeting, Eisenhower asked for a JCS study on the mobilization base to be "done by young, imaginative majors instead of '———'!" CNO Burke to General Picher, memorandum, July 23, 1959, in CCS 381 folder, CJCS Records, RG 218, NA.

49. Cutler to Herter, letter, January 27, 1958, *DDC*, 1989/2985.

50. "Discussion Paper on U.S. Policy in the Event of War Initiated by the Sino-Soviet Bloc," enclosure to memorandum, James Lay to the NSC, January 7, 1959, "Subject: Review of NSC 5410/1," *DDQC*, 1978/60B.

51. Memorandum, "Subject: Discussion at the 394th Meeting of the NSC, Thursday, January 22, 1959," January 22, 1959, *DDC*, 1990/1010.

52. National Security Council, "U.S. Policy in the Event of War," NSC 5904, February 19, 1959, *DDC*, 1986/484.

53. Stephen E. Ambrose, *Eisenhower*, vol. 2: *The President* (New York: Simon & Schuster, 1984), pp. 515–520. See also John Eisenhower, MCP [on March 6, 1959, 10:30 A.M.], March 6, 1959, *DDC*, 1992/2764.

54. Memorandum, "Subject: Discussion at the 398th Meeting of the NSC, Thursday, March 5, 1959," March 5, 1959, *DDC*, 1990/1014. The remainder of the meeting was devoted largely to editorial issues of minor policy import.

55. James Lay to the NSC, memorandum, March 9, 1959, "Subject: NSC 5904" and attachment to memorandum, "Subject: Discussion at the 399th Meeting of the NSC, Thursday, March 12, 1959," *DDC*, 1990/1018.

56. Twining to Gates, memorandum, April 11, 1960, "Subject: U.S. Policy in the Event of War," in CCS 381 folder, CJCS Records, RG 218, MMB, National Archives.

57. See Richard Betts, *Nuclear Blackmail and Nuclear Balance* (Washington, D.C.: Brookings Institution, 1987). After the 1958 Taiwan Straits crisis, General Laurence Kuter, the Air Force commander in the Pacific, briefed an Air Force commanders conference. He told them: "I think probably the most important lesson we have learned from Taiwan is that we must have a firm policy regarding the use of nuclear weapons. A priority requirement is to educate our various government policy

makers that the very great spread available in nuclear weapons has made these weapons conventional." Laurence Kuter, "Report on Taiwan Straits Situation," Speech series, Laurence Kuter Papers, Air Force Academy Library.

58. Pruessen writes that Eisenhower and Dulles behavior was "symptomatic of a certain cockiness about control capabilities as well, a sometimes frighteningly naive assumption that the decision to unsheathe weapons of mass destruction would always be one's own and would only be taken after mature deliberation." Ronald W. Pruessen, "John Foster Dulles and the Predicaments of Power," in Immerman, *John Foster Dulles*, p. 36.

59. Attachment to letter, Foster to Killian, January 15, 1958.

60. Wentworth to the JCS, memorandum, SM-12-58, January 4, 1958; and Army Flimsy, "Subject: JCS 1844/242—Atomic Annex to JSCP [Joint Strategic Capabilities Plan]," January 7, 1958. In CCS 373.11 folder, JCS Records, RG 218, MMB, NA. See also David Alan Rosenberg, "The Origins of Overkill: Nuclear Weapons and American Strategy, 1945–1960," *International Security* 7 (1983):53.

61. Joint Secretariat to the JCS, note, December 30, 1957, "The Character and Probable Results of a Future General War."

62. Morse to Cutler, memorandum, March 8, 1958, "Subject: Massive Deterrent," enclosure to letter, Morse to Cutler, March 22, 1958.

63. The zeros probably indicate the actual numbers that Cutler added orally at the briefing. Robert Cutler, "Massive Exchange of Nuclear Weapons," paper, March 16, 1958, *DDQC*, 1991/2647.

64. Cutler to Twining, memorandum, March 24, 1958, "Subject: Enemy Targets to be used in the 1958 Net Evaluation," supplied in response to author's FOIA request.

65. Cutler added: "'Military requirements' for nuclear weapons and forces should not exceed the possible and most effective use of weapons and forces made available—at vast expense—to meet such 'requirements'." Cutler, March 16, 1958, "Massive Exchange of Nuclear Weapons." Also Cutler to Eisenhower, memorandum, March 18, 1958, "Subject: November, 1958, Net Evaluation," in Nuclear Policy (1958) folder, NSC series, Briefing Notes subseries, WHO-OSANSA, DDEL.

66. Cutler, Memorandum for Record, May 21, 1958, in Missiles and Military Space Programs 1955–1961 (2) folder, NSC series, Briefing Notes subseries, WHO-OSANSA, DDEL.

67. The report stated that "for purposes of deterrence, or of general war if deterrence should fail, no one type of weapons system would suffice. The requirements imposed by the situation and by the target systems are such that the best choice of systems is a balanced combination of manned aircraft and ballistic and/or aerodynamic missiles. This balance offers the possibilities of different weapons systems complementing each other, re-enforcing each other, complicating the enemy's defenses, and reducing the chance of his destroying most of our striking power by surprise attack." Weapons Systems Evaluation Group, "First Annual Review of WSEG Report no. 23—The Relative Military Advantages of Missiles and Manned Aircraft (hereafter cited as WSEG-23/1)," August 8, 1958, p. 2, in Targeting and "The Relative Military Advantages of Missiles and Manned Aircraft" folder, NSC series, Briefing Notes subseries, WHO-OSANSA, DDEL.

68. According to the WSEG, "the Bravo target system is intended to meet the requirement stated under alternative undertakings in the atomic annex of the current Joint Strategic Capabilities Plan: 'To provide for general war initiated under

disadvantageous conditions wherein the available U.S. forces are considered insufficient due to enemy action or otherwise, for the accomplishment of the (Primary Undertakings)'." Ibid., p. 9.

69. "The earlier advent of the Soviet ICBM's offers a more difficult counterforce problem which may be met through our development of a more accurate, faster-reacting, ballistic missiles, provided definite intelligence of ICBM site locations is obtained." Ibid., p. 5.

70. "The dual nature of our military objectives for general war—deterrence through a retaliatory capability and an effective counterforce capability—does not require that we possess separate weapons for each purpose. Weapons suited to a counterforce role possess, in general, a significant capability for pure retaliation and hence are effective for purposes of deterrence. The converse is not necessarily true." Ibid., p. 19.

71. Gray, MMP [on August 26, 1958], August 26, 1958, *DDQC*, 1985/1429.

72. Twining to Gray, letter, September 5, 1958, in Missiles and Military Space Programs, 1955–1961 (2) folder, NSC series, Briefing Notes subseries, WHO-OSANSA, DDEL. See also Gray to Twining, letter, August 29, 1958, *DDQC*, 1982/2882.

73. "Presentation by the Director, WSEG, to the National Security Council on the Subject of Offensive and Defensive Weapons Systems," October 13, 1958, supplied in response to author's FOIA request.

74. NSC Action no. 1994*b*, "Record of Actions by the NSC at Its 382nd Meeting Held on October 13, 1958," *DDQC*, 1985/1817; and memorandum, "Subject: Discussion at the 382nd (Special) Meeting of the NSC, Monday, October 13, 1958," *DDC*, 1990/357.

75. Gerhart wrote: "I believe this is not simply because of the tremendous casualties which would result but specifically because of the destruction of government controls and civil disruption which should be achieved." Gerhart to Gray, letter, October 13, 1958, in Targeting and "The Relative Military Advantages of Missiles and Manned Aircraft" folder, NSC series, Briefing Notes subseries, WHO-OSANSA, DDEL.

76. Ibid.

77. Gray, MCP [on October 16, 1958, at 11:00 A.M.], October 17, 1958, *DDC*, 1986/585; Gray to Twining, letter, October 30, 1958, *DDQC*, 1981/607A, and Killian to Gray, memorandum, October 29, 1958, supplied in response to author's FOIA request.

78. "We [can] not separate the question of effective deterrence from the question of war objectives and neither could be separated from our targeting plans." Gray, MCP [on Wednesday, November 19, 1958, at approximately 3:45 P.M.], November 22, 1958, in Meetings with the President, 1958 (1) folder, Special Assistants series, Presidential subseries, WHO-OSANSA, DDEL.

79. Memorandum, "Subject: Discussion at the 387th Meeting of the NSC, November 20, 1958," November 20, 1958, *DDC*, 1992/2737.

80. NSC Action no. 2009*b*, "Record of Action by the NSC at Its 387th Meeting Held on November 20, 1958," in Record of Actions by NSC 1958 (3) folder, NSC series, ACW file, DDEL.

81. Gleason to Gray, memorandum, January, 7, 1959; and Gray to Gerhart, letter, January 12, 1959. In Target System 1957–1960 folder, NSC series, Briefing Notes subseries, WHO-OSANSA, DDEL.

82. Gray asked McElroy:

"c. Is there a valid requirement to develop another generation of strategic bombers, taking into account the probable capabilities of enemy defenses and the problems of maintaining such aircraft in an alert status?

"d. Would it be desirable to extend the effectiveness of Polaris missiles by adapting them for use on surface vessels, and on land bases, assuming the development of an appropriate mobile launching unit?

"e. Should further priority be given to Minuteman by the earlier establishment of a production capability and by seeking to afford it more mobility? What effect would further emphasis on Minuteman have on the Titan and Atlas missiles program?

"f. What is the proper balance between procurement of additional retaliatory weapons and expenditures to protect those now available against surprise attack?"

Gray to McElroy, memorandum, February 18, 1959, "Subject: Defense Presentations to the President," *DDQC*, 1981/607B.

83. Twining also directed the NESC to determine "the adequacy of the required retaliatory forces [in each target system] to contribute to the national objective of deterrence." "Appraisal of Relative Merits, from the Point of View of Effective Deterrence, of Alternative Retaliatory Efforts," February 18, 1959, enclosure to memorandum, Twining to Gray, February 19, 1959, "Subject: NSC Action no. 2009," in Defense Presentations to NSC, 1959–1960 folder, NSC series, Briefing Notes subseries, WHO-OSANSA, DDEL.

84. "Enemy defense suppression" refers to the number of weapons targeted against Soviet air defense sites. "Enemy resistance survival" is "the fraction of the launched force penetrating the line of enemy resistance and surviving to the Bomb Release line." Ibid.

85. J-5 to the JCS, Report on Weapons Systems for Strategic Delivery, January 14, 1959, JCS 1620/215, in CCS 4600 folder, JCS Records, RG 218, MMB, National Archives.

86. CSUSAF to the JCS, Memorandum on Weapons Systems for Strategic Delivery, January 18, 1959, JCS 1620/216, in CCS 4600 folder, JCS Records, RG 218, MMB, National Archives.

87. CJCS Twining to McElroy, memorandum, January 20, 1959, "Subject: Weapons Systems for Strategic Delivery," in CCS 4600 folder, JCS Records, RG 218, MMB, National Archives.

88. White to Kuter, letter, March 3, 1959; and White to Power, letter, May 11, 1959. In Top Secret General (1959) folder, White Papers, Library of Congress. In a top secret lecture, Lt. General Gerhart called minimum deterrence an "extremely dangerous" nuclear strategy. Gerhart, "An Air Force View of the Current Military Strategy of the United States for Cold, Limited, and General War," presentation, Air Force Historical Research Center (AFHRC), Maxwell Air Force Base (MAFB), Ala.

89. J-5 to the JCS, Report on the Evaluation of Offensive and Defensive Weapons Systems, June 23, 1959, JCS 1620/263, in CCS 4600 folder, JCS Records, RG 218, MMB, National Archives.

90. "By the end of this period, with a continuance of present trends and programs on both sides, and with no major technological breakthroughs on either side in the intervening years, the most probable position will be that of each side having military strength of potentially decisive proportions with an advantage, possibly conclusive, to the side taking the initiative." Burke to McElroy, memorandum, July 13, 1959, "Subject: Relative Military Capabilities in the 1959–1961/62 Time Period," in JCS 1(7) folder, Subject series, DOD subseries, White House Office—Office of the Staff Secretary (WHO-OSS), DDEL.

91. LeMay to the JCS, memorandum, June 3, 1958, "Subject: JSOP [Joint Strategic Objectives Plan] for 1 July 1962," in CCS 381 folder, JCS Records, RG 218, MMB, National Archives.

92. Power to Twining, memorandum, March 6, 1959, enclosure to note on Polaris Weapon System, Secretaries to the JCS, April 3, 1959, JCS 1620/242, in CCS 4270 folder, JCS Records, RG 218, MMB, National Archives.

93. White to the JCS, Memorandum on Command and Control of Strategic Forces, April 28, 1959, JCS 1620/250, in April 1959 folder, Nuclear History series, National Security Archive.

94. Burke to the JCS, Memorandum on Concept of Employment and Command Structure for the Polaris Weapon System, April 30, 1959, JCS 1620/254, in CCS 4270 folder, JCS Records, RG 218, MMB, National Archives.

95. Burke to McElroy, memorandum, May 5, 1959, "Subject: Statement of Navy Views on the Concept of Employment and Command Structure for Polaris Weapon System," in CCS 4270 folder, JCS Records, RG 218, MMB, National Archives.

96. Appendix to memorandum, Twining to the JCS, August 24, 1959, "Subject: Target Coordination and Associated Problems," in August 1959 folder, Nuclear History series, National Security Archive.

97. Gray, MMP [on January 13, 1960], January 15, 1960, *DDQC*, 1985/2144.

98. "SIOP-62 Briefing," *International Security* 12 (1987):43.

99. Ibid. In "The Origins of Overkill" (p. 62) Rosenberg states that this target list included "121 ICBM sites, 140 air defense bases, 200 bomber bases, 218 military and governmental centers, and 124 other military targets . . . with the remaining targets apparently located primarily within 131 urban centers."

100. Gray, MMP [on Wednesday, February 17, 1960 at 3:00 P.M.], February 19, 1960, *DDQC*, 1985/2146.

101. Goodpaster, MCP [on February 12, 1960], February 18, 1960, in JCS 1(8) (September 1959–May 1960) folder, Subject series, DOD subseries, WHO-OSS, DDEL.

102. Goodpaster, MCP [on June 24, 1959], June 24, 1959, *DDQC*, 1978/312C; and report, "Effects of Fallout in Nuclear War," in Exchange of Nuclear Stockpiles folder, Executive Secretary series, WHO-NSC, DDEL.

103. Goodpaster, MCP [on May 5, 1960], May 7, 1960, in JCS 1(8) folder, Subject series, DOD subseries, WHO-OSS, DDEL.

104. White to Gates, letter, June 10, 1960, in June 1960 file, Nuclear History series, National Security Archive. For an alternative proposal, see "Concept for Integrated Strategic Operations," attachment to memorandum, Evan Aurand to Kistiakowsky, July 11, 1960, *DDQC*, 1981/609B.

105. See White to Power, letter, July 19, 1960, "Recommended Actions for Increasing Capabilities and Readiness Posture of SAC Forces," in July 1960 folder, Nuclear History series, National Security Archive.

106. John Eisenhower, MCP [on July 6, 1960, at 12:30 P.M.], July 6, 1960, *DDQC*, 1980/108B.

107. Gates to CJCS, Draft Memorandum, "Subject: Target Coordination and Associated Problems," *DDQC*, 1981/287B; Gates to Goodpaster, letter, August 10, 1960, in DOD (Aug. 1960) 4(7) folder, Subject series, DOD subseries, WHO-OSS, DDEL.

108. Eisenhower believed the directive "should not be given such categorical approval on his part that a later decision to discontinue it would destroy confidence in the soundness of his judgment and approval in such things." Goodpaster, MCP

[on August 11, 1960], August 13, 1960, *DDQC*, 1982/37; Burke to Eisenhower, Letter, August 11, 1960, *DDQC*, 1981/379A.

109. Goodpaster, MCP [on October 11, 1960], October 12, 1960, in Staff Notes, October 1960 (1) folder, DDE Diary series, ACW file, DDEL. See also memorandum, "Subject: Discussion at the 462nd Meeting of the NSC Thursday, October 6, 1960," October 6, 1960, *DDC*, 1991/2032.

110. Ibid. On the interservice politics involved in SIOP formulation, see Lieutenant General Glenn A. Kent Oral History, pp. 39–43, AFHRC, MAFB.

111. Chart, "SIOP—1 April 1961," enclosure to memorandum, Cyrus Vance to President Johnson, October 3, 1964, "Subject: Military Strength Increases since Fiscal Year 1961," *DDQC*, 1978/350A.

112. "SIOP-62 Briefing," pp. 49–50.

113. Eisenhower to Kistiakowsky, memorandum, October 19, 1960, in DOD (1960) (7) folder, WHO-OSST, DDEL.

114. Kistiakowsky to Goodpaster, memorandum, November 7, 1960, in Kistiakowsky (6) (October–December 1960) folder, Subject series, Alph Subseries, WHO-OSS, DDEL.

115. Goodpaster, MCP [on November 25, 1960], December 1, 1960, *DDC*, 1988/1185.

116. According to the directorate, the five targeting tasks were to stop nuclear strikes against the United States, to stop nuclear strikes against the Allies, to retard conventional operations, to disrupt enemy war-making capacity, and to gain control of enemy air space. Attachment to memorandum, General LeMay to General White, March 2, 1960, "Subject: The Threat," in March 1960 folder, Nuclear History series, National Security Archive.

117. Ginsburgh to White, memorandum, October 21, 1960, "Subject: Counterforce Studies"; Richardson to Gerhart, memorandum, November 9, 1960; Ginsburgh to Gerhart, memorandum, November 17, 1960; and White to Gerhart, memorandum, December 30, 1960. In Chief of Staff memos folder, White Papers, Library of Congress.

118. The WSEG informed Rathjens of its conclusions, however. George Rathjens, Memorandum for Record, September 21, 1960, "Subject: The General Balance of Strategic Forces and Their Relation to the FY 1962 Budget," *DDC*, 1988/561. Gates sought access to the sections of the study on command and control. See Gray to Gates, letter, November 5, 1960, in November 1960 folder, Nuclear History series, National Security Archive; and Gates to Twining, memorandum, November 14, 1960, "Subject: Survival of Decision Making Machinery," in CCS 4600 folder, JCS Records, RG 218, MMB, National Archives.

119. Desmond Ball, *Politics and Force Levels* (Berkeley: University of California Press, 1980), pp. 34–38.

120. The weapons included in the study were the B-47, B-52, B-58, B-70, Skybolt, Hound Dog, Atlas, Titan, Polaris, and both configurations of Minuteman. The characteristics examined were "reliability, reaction time, responsiveness to control, penetration capability, accuracy, destructiveness, vulnerability and cost." WSEG, "Evaluation of Strategic Offensive Weapons Systems," Report no. 50 (WSEG-50), December 27, 1960, in December 1960 folder, Nuclear History series, National Security Archive.

121. "The Feasibility of Achievement of Counterforce Objectives," appendix E to enclosure A to ibid., pp. 70–71.

122. See the charts accompanying ibid. According to most estimates, the point at

which United States weapons deployment would begin to have diminishing returns was between 800 weapons and 1,200 weapons, with virtually no value expected beyond the deployment of 1,400.

123. Ibid., p. 80.

124. "Strategic Implications of Possible Changes in the Nature of the Threat," enclosure J to WSEG-50, in CCS 4600 folder, JCS Records, RG 218, MMB, National Archives.

125. Appendix E to enclosure A to WSEG-50, p. 82.

126. "The safest way to give evidence of our own restraint will be to limit the number of issues on which strategic sanctions are threatened. . . . [Avoiding accidental war] requires impressing the enemy with the proposition that he will avoid serious dangers by observing the restraints that our own moves may suggest. Such an impression may depend on Soviet knowledge that the U.S. possesses sufficient graduated forms of military power to significantly widen the scope of 'local' conflicts should it choose to do so, without going all the way to an unrestricted, uncontrolled thermonuclear exchange." Enclosure J to WSEG-50, pp. 3, 16, 21.

127. WSEG-50, p. 24.

128. Rathjens, Memorandum for Record, September 21, 1960.

129. Christian Herter, memorandum, November 8, 1960, *DDC*, 1991/2642.

4. Defense Budgets and the Public Debate over the Missile Gap

1. Memorandum, "Subject: Discussion at the 317th Meeting of the NSC, March 28, 1957," March 29, 1957, *FRUS*, 1955–1957, vol. 19: *National Security Policy* (Washington, D.C.: GPO, 1990), pp. 446–456.

2. John W. Sloan, *Eisenhower and the Management of Prosperity* (Lawrence: University Press of Kansas, 1991), pp. 143–144.

3. The Eisenhower administration created the Gaither panel in response to this report. See memorandum, "Subject: Discussion at the 318th meeting of the NSC, April 4, 1957," *FRUS*, 1955–1957, 19:459–464.

4. Goodpaster, MCP [on May 20, 1957], May 21, 1957, ibid., pp. 486–488.

5. Memorandum, "Subject: Discussion at the 319th meeting of the NSC, April 11, 1957," April 12, 1957, and Memorandum, "Subject: Discussion at the 320th meeting of the NSC, April 17, 1957," April 19, 157, ibid., pp. 470, 480–486.

6. See Goodpaster, MCP [on May 20, 1957], May 21, 1957, ibid.

7. Wilson to Eisenhower, memorandum, July 10, 1957, ibid., p. 545.

8. Goodpaster, MCP [on July 10, 1957, at 10:00 A.M.], July 12, 1957, ibid., pp. 547–548.

9. Memorandum, "Subject Discussion at the 331st Meeting of the NSC, July 18, 1957," July 19, 1957, ibid., p. 555.

10. Radford to Wilson, memorandum, July 16, 1957, ibid., pp. 549–553.

11. Memorandum, "Subject: Discussion at the 332nd Meeting of the NSC, July 25, 1957," July 26, 1957, ibid., pp. 556–565.

12. Ibid., p. 560, emphasis added.

13. Memorandum, "Subject: Discussion at the 332nd Meeting of the NSC, July 25, 1957," ibid., pp. 556–565.

14. Memorandum, "Subject: Discussion at the 329th Meeting of the NSC, July 3, 1957," ibid., p. 537.

15. Memorandum, "Subject: Discussion at the 333rd Meeting of the NSC, August 1, 1957"; and Wilson to Eisenhower, letter, August 9, 1957. In ibid., pp. 568–571, 580–583.

16. See "Democratic Fact Sheet," October 24, 1957, in National Defense—Missiles (October 24, 1957–January 3, 1958) folder, Legislative Assistant Background file (1953–1960), John F. Kennedy Pre-presidential Papers, JFKL.

17. An article in the *New York Times* was critical of the administration: "The military missiles program has been subjected to recent curtailments in expenditures and developments, and the satellite program, conducted at bargain-basement rates as a purely scientific experiment, reveals an unexpected naivete in Washington." A *Denver Post* editorial blamed the president himself: "COME OFF IT, IKE, WHO YOU KIDDING? President Eisenhower is the only person we know who appears to be undisturbed about Sputnik. And that worries us most." See "Democratic Fact Sheet," November 1, 1957, in National Defense—Missiles (October 24, 1957–January 3, 1958) folder, Legislative Assistant Background file (1953–1960), John F. Kennedy Pre-presidential Papers, JFKL.

18. Cutler, note, October 23, 1957, *DDC*, 1986/1085.

19. Walter A. McDougall, . . . *The Heavens and the Earth: A Political History of the Space Age* (New York: Basic Books, 1985), p. 150.

20. See Chapter 1. The panel recommended allocating $19 billion for offensive forces and $25 billion for passive measures over the next five years. This would require an initial increase in the FY 1959 defense budget of $4.73 billion and the highest increase, in FY 1961, of $11.92 billion. See the Gaither report.

21. Memorandum, "Subject: Discussion at the 343rd Meeting of the NSC, November 7, 1957," in 343rd meeting of the NSC on November 7, 1957 folder, NSC series, ACW file, DDEL.

22. Cutler informed the panel of the administration's desires on at least two occasions. "The last thing the President wants to come out of this study," Cutler wrote, "is a series of detailed recommendations for changing our defense programs." Cutler, note, June 20, 1957, *DDQC*, 1982/2802.

23. Eisenhower first indicated that he might approve defense spending that exceeded $38 billion in a meeting with McElroy on October 14. Goodpaster recorded that "McElroy referred to the $38 billion figure and said he would like not to regard it as a rigid ceiling. The President welcomed this comment, saying that he had never wanted it to be called that. As a matter of fact, *he had not wished to establish a figure at all, but had done so on the repeated request of Secretary Wilson.* Although he had set it as a judgment as to a proper level, he promptly found that it was being spoken of as a rigid ceiling" (emphasis added). Eisenhower's attempt to disavow his role in setting the budget ceiling is not supported in the documentary record. The president did nothing to dissuade Wilson from conceptualizing the budget plan as a ceiling, despite numerous opportunities to do so in the previous spring and summer. In fact, it would seem that the president encouraged Wilson to do so in the hope of achieving long-term budget discipline within the DOD. Goodpaster, MCP [on October 14, 1957], October 15, 1957, *FRUS, 1955–1957*, 19:605.

24. Goodpaster, MCP [on November 11, 1957, at 5:00 P.M.], November 16, 1957, in Military Budget (6) folder, Subject series, DOD subseries, WHO-OSS DDEL.

25. Memorandum, "Subject: Discussion at the 345th Meeting of the NSC on Thursday, November 14, 1957," November 15, 1957, *DDC*, 1987/1010.

26. Cutler, notes, November 14, 1957, in 345th NSC Meeting, on November 14, 1957, folder, NSC series, ACW file, DDEL.

27. Memorandum, "Subject: Discussion at the 346th Meeting of the NSC on Friday, November 22, 1957," November 25, 1957, *DDC*, 1987/1690, p. 2.

28. Goodpaster, MCP [on December 5, 1957], December 5, 1957, *FRUS*, 1955–1957, 19:703.

29. See C. D. Jackson log, October 14–November 11, 1957, *DDQC*, 1977/155B.

30. Memorandum, "Bipartisan Congressional Meeting," December 3, 1957, Legislative Meetings series, ACW file, DDEL.

31. See Morton Halperin, "The Gaither Committee and the Policy Process," *World Politics* 13 (April 1961):360–384; and Fred Kaplan, *The Wizards of Armageddon* (New York: Simon & Schuster, 1983), pp. 152–154.

32. Nitze complained that Dulles's criticism had undermined the impact of the Gaither report on Eisenhower. Nitze to John Foster Dulles, letter, November 16, 1957, in November 1957 folder, Nuclear History series, National Security Archive. Dulles, incredulous when he received Nitze's letter suggesting that he resign, called his brother, Allen, to find out how Nitze had become involved in the Gaither report. John Foster Dulles to Allen Dulles, telephone call, Monday, November 18, 1957, at 6:44 P.M., ibid.

33. After hearing the presentation of the Gaither report to the NSC on November 7, Eisenhower remarked, that "it would be interesting to make a test to find out how long the matters which had come before the Council this morning could be kept secret." Memorandum, "Subject: Discussion at the 343rd Meeting of the NSC, Thursday, November 7, 1957." See also CNO to the JCS, memorandum, December 20, 1957, "Leak of Gaither Report to Press," JCS, 1810/70, *DDQC*, 1983/207.

34. Nixon and Gordon Gray supported a sanitized release since it would be less sensational than news reports, but Cutler voiced "violent opposition." Memorandum, "Subject: Discussion at the 350th Meeting of the NSC, January 6, 1958," January 7, 1958, *DDC*, 1990/301. The White House drafted a letter to H. Rowan Gaither and a press release but neither was ever sent. See "Possible Draft Letter" and "Possible Draft Statement," January 9, 1958, *DDQC*, 1983/523, 524.

35. Perkins to Killian, letter, January 9, 1958, in Security Resources Panel folder, WHO-OSST, DDEL. See also, Gaither to Killian, letter, January 14, 1958, in the same folder. E. P. Oliver, Rand Corporation associate and technical adviser to the panel, instructed members "to take every advantage of this favorable environment in order to impress the highest levels of Government and all concerned with the unanimity of their belief in at least the highest priority category of recommendations." Oliver to the Security Resources Panel Steering and Advisory Panel, memorandum, January 14, 1958, *DDQC*, 1981/512A.

36. See Stephen E. Ambrose, *Eisenhower*, vol. 2: *The President* (New York: Simon & Schuster, 1984), pp. 434–435; and Gregg Herken, *Counsels of War* (New York: Alfred A. Knopf, 1985), pp. 116–117.

37. This section reviews the budgetary aspect of the discussions while the next chapter examines the programmatic aspects.

38. JCS to McElroy, draft memorandum, "Subject: Augmentation of the Present ICBM and Polaris/Submarine Weapons Systems," enclosure A to Joint Strategic Plans Committee to the JCS, March 26, 1958, "Report to the President by the Security Resources Panel of the ODM Science Advisory Committee (The Augmentation of the Present ICBM and Polaris/Submarine Weapons Systems)," JCS 2101/302, *DDQC*, 1981/158A. See also CJCS to the JCS, memorandum, February 5, 1958, "Evaluation of Offensive and Defensive Weapons Systems," JCS 1620/174, *DDQC*, 1983/173.

39. The CSUSA and the CNO recommended no additional funding for Titan, but the CSUSAF and the Marine Corps commandant endorsed the additional $100 million. With regard to Polaris, the CNO and the Marine Corps commandant supported the $400 million figure, and the CSUSAF and the CSUSA sought only $83 million for solid fuel research and development. DDQC, 1981/158A.

40. Goodpaster, MCP [on March 20, 1958, at 3:00 P.M.], March 21, 1958, in Missiles and Satellites 2(3) (March 16, 1958–June 30, 1958) folder, Subject series, DOD subseries, WHO-OSS, DDEL.

41. McElroy's memo for the meeting outlining the program is available as "Preliminary Department of Defense Proposals for '59 Budget Supplemental Appropriations," *DDQC*, 1983/163.

42. Goodpaster, MCP [on March 20, 1958, at 4:00 P.M.], March 21, 1958, Missiles and Satellites 2(3) (March 16, 1958–June 30, 1958) folder, Subject series, DOD subseries, WHO-OSS, DDEL.

43. Memorandum, "Subject: Discussion at the 363rd Meeting of the NSC, Thursday, April 24, 1958," April 25, 1958, *DDQC*, 1980/384B.

44. Goodpaster recorded: "The President said that our country can afford what is needed for defense, but it cannot afford the costs that might be added by wastefulness." Goodpaster, MCP [on June 23, 1958], June 23, 1958, in JCS 1(4) folder, Subject series, DOD subseries, WHO-OSS, DDEL

45. Two NSC staffers [Robert Johnson and George Weber], favored a review by the service secretaries rather than by the NSC Planning Board for strikingly Allisonian reasons: "If the Planning Board as a whole prepares the initial report, the pulling and hauling between Budget-Treasury on one side and the agencies responsible for the programs on the other tends to result in meaningless statements of issues or recommendations." The opposition of Budget Bureau Director Stans prevented an initial NSC review of budget priorities. See Johnson to Weber, memorandum, April 25, 1958, "Relation of NSC to the Budget-Making Process"; and Johnson to Cutler, memorandum, May 19, 1958, "Subject: National Security and the Budget." In Budget—Federal (3) (February 1956–July 1958) folder, Special Assistant series, Subject subseries, WHO-OSANSA, DDEL.

46. "Statement of Assistant Secretary of Defense W. J. McNeil before the Cabinet," July 25, 1958, attachment to memorandum, R. Eugene Livesay to Gray, August 4, 1958, *DDC*, 1990/1869.

47. Stans to McElroy, letter, October 27, 1958, in DOD (5) (September–December 1958) folder, WHO-OSS, DDEL. See also memorandum, "Major Decisions Involved in Processing the FY 1960 Budget for the Department of Defense," October 28, 1958, in DOD (15) folder, Subject series, Confidential file, DDEL, p. 4.

48. Memorandum, "Major Decisions Involved in Processing the FY 1960 Budget for the Department of Defense," October 28, 1958.

49. In a memorandum, the OSST asked: "Is the size of the strategic striking force, therefore, to be determined principally in terms of its expected effectiveness against bases of the enemy striking force, or by its effectiveness for retaliatory purposes, or both? If intended targets consist of population centers and enemy war mobilization potential, would a smaller force be possible? Alternatively, if intended targets consist of enemy striking force bases alone, or such bases together with population centers, will a larger force be necessary?" Memorandum, OSST, "Selected Issues in the FY '60 Defense Budget (as of November 1, 1958)," in DOD folder (September–December 1958), WHO-OSST, DDEL.

50. Memorandum, "Major Decisions Involved in Processing the FY 1960 Budget for the Department of Defense," October 28, 1958.

51. John Foster Dulles, Memorandum of Conversation with Secretary of the Treasury Robert Anderson, November 6, 1958, *DDC*, 1986/807; John Foster Dulles, Memorandum of Conversation, Wednesday, November 26, 1958, at 5:15 P.M., in November 1958 folder, Nuclear History series, National Security Archive; and Gray, Memorandum of Conference, with Secretary Dulles, December 3, 1958, in Memoranda for Record (1) folder, Special Assistant series, Subject subseries, WHO-OSANSA, DDEL.

52. Bureau of the Budget Military Division to Stans, memorandum, November 27, 1958, "Fiscal Year 1960 Budget—DOD," in Offensive and Defensive Weapons Systems and the FY '60 Defense Budget (2) folder, NSC series, Briefing Notes subseries, WHO-OSANSA, DDEL.

53. John Eisenhower, MCP [on November 28, 1958, at 8:45 A.M.], December 9, 1958, *DDC*, 1992/520.

54. Memorandum, BoB Military Division to Stans, November 27, 1958, "Fiscal Year 1960 Budget—DoD."

55. John Eisenhower, MCP [on November 28, 1958, at 8:45 A.M.], December 9, 1958, *DDC*, 1992/520.

56. "Program Decisions Which Should Be Made," December 3, 1958, *DDC*, 1991/2457.

57. "[T]he Joint Chiefs of Staff consider that the FY 1960 proposed expenditure figure of $41,165,000,000 is adequate to provide for the essential programs necessary for the defense of the period under consideration. They find no serious gaps in the key elements of the budget in its present form, but all have reservations with respect to the funding of some segments of their respective service programs." Lieutenant Colonel Camm, memorandum, January 7, 1959, *DDC*, 1986/82, 83.

58. Memorandum, "Subject: Discussion at the 389th Meeting of the NSC, December 6, 1958," *DDC*, 1991/1994; NSC Action no. 2013, "Record of Actions by the NSC at Its 389th Meeting Held on December 6, 1958," in Record of Actions by NSC 1958 (3) folder, NSC series, ACW file, DDEL.

59. Gray, MCP [December, 8, 1958, at 11:15 A.M.], December 9, 1958, *DDQC*, 1985/2138.

60. Several good treatments of the missile gap are: Edgar Bottome, *The Missile Gap: A Study of the Formulation of Military and Political Policy* (Teaneck, N.J.: Fairleigh Dickinson University Press, 1971); John Prados, *The Soviet Estimate* (New York: Dial Press, 1982); and Kaplan, *The Wizards of Armageddon*, chap. 10.

61. The WSEG acknowledged this danger in WSEG-23/1.

62. L. A. Minnich, memorandum, January 5, 1959, "Bipartisan Congressional Meeting," *DDQC*, 1976/299C.

63. Quoted in Felix Belair, Jr., "President to Seek 41 Billion in Arms," *New York Times*, January 6, 1959, p. 1.

64. Albert Wohlstetter, "The Delicate Balance of Terror," *Foreign Affairs* 37 (1959):209–234.

65. "President Notes Missile Progress," *New York Times*, January 15, 1959, p. 1.

66. "Presentation by the Chairman, Joint Chiefs of Staff," *DDC*, 1989/1936.

67. U.S. Congress, Senate Committee on Foreign Relations, *Executive Sessions of the Senate Foreign Relations Committee* (historical series), vol. 11 (1959), 86th Cong. 1st sess. (Washington, D.C.: USGPO, 1982), pp. 18–19.

68. Ibid., pp. 33–44 (emphasis added).

69. "M'Elroy Says U.S. Won't Race Soviet in Making ICBM's," *New York Times*, January 30, 1959, p. 1.

70. Ibid.

71. Walter Barton Leach to John F. Kennedy, letter, January 16, 1959, Walter Barton Leach Papers, Harvard Law School Library.

72. Symington to Twining, letter, February 18, 1959; and Twining to Symington, letter, February 20, 1959. In CJCS 471.94 folder, Twining CJCS Records, RG 218, MMB, NA.

73. John Eisenhower, MCP [on February 9, 1959, at 10:30 A.M.], February 10, 1959, in JCS 1(6) folder, Subject series, DOD subseries, WHO-OSS, DDEL; and Goodpaster, MCP [on February 10, 1959], February 13, 1959, *DDC*, 1987/464.

74. Quoted in Jack Raymond "Air Force Chief Backs President on Defense Issue," *New York Times*, March 16, 1959, p. 1. On the hearings, see Edward Kolodziej, *The Uncommon Defense and Congress, 1945–1963* (Columbus: Ohio State University Press, 1966), pp. 292–298.

75. Memorandum, "Subject: Discussion at the 408th Meeting of the NSC, May 28, 1959," *DDC*, 1990/851.

76. See Evan Aurand to Eisenhower, memorandum, April 21, 1959, *DDQC*, 1978/448C; and Goodpaster, MCP [on February 10, 1959], February 13, 1959, *DDC*, 1987/464.

77. Some congressional leaders were also hesitant to "interfere" with executive prerogatives. John Eisenhower, MCP [on February 9, 1959, at 10:30 A.M.], February 10, 1959.

78. Kolodziej, *The Uncommon Defense and Congress*, p. 310.

79. Maxwell Taylor, *The Uncertain Trumpet* (New York: Harper & Brothers, 1960), p. 92.

80. This section is based on Alfred Goldberg, *Strategy and Money: The Defense Budget for Fiscal Year 1961* (Washington, D.C.: USAF Historical Division, July 1963), AFHRC, MAFB. This study was classified "secret" but was declassified in June 1985.

81. LeMay to Secretary James Douglas, letter, September 9, 1959, in 1959–Sec. AF folder, White Papers, Library of Congress.

82. Goodpaster, MCP [on November 3, 1959], November 6, 1959, in DOD 3(9) folder, Subject series, DOD subseries, WHO-OSS, DDEL, and Goodpaster, MCP [on November 5, 1959], November 6, 1959, in Staff Notes—November 1959 (3) folder, DDE Diary series, ACW file, DDEL.

83. Goodpaster, MCP [on Monday, November 16, 1959, at 8:30 A.M. in Augusta, Georgia], December 2, 1959, *DDQC*, 1980/107D.

84. Goodpaster, MCP [on November 18, 1959, in Augusta], January 20, 1960, *DDQC*, 1981/608.

85. Robert Merriam, Memorandum for Record, November 21, 1959, in Staff Notes—November 1959 (2) folder, DDE Diary series, ACW file, DDEL.

86. Goodpaster recorded that the following telling exchange took place at the meeting: "The President commented that the capabilities of missiles for destruction are beyond human comprehension. It is very difficult to plan, because our plans are developed out of past experience. Dr. [Herbert] York commented that the total deliverable destructive power is the significant thing and the President agreed." Goodpaster, MCP [on November 21, 1959, in Augusta], January 2, 1960, *DDQC*, 1980/108A.

87. Bernard Nalty concluded that these "were mere shells of the complex weapon

system sought by the Air Force." Bernard Nalty, *The Quest for an Advanced Manned Strategic Bomber: USAF Plans and Policies, 1961–1966* (Washington, D.C.: USAF Historical Division, 1966), p. 2, AFHRC, MAFB.

88. Memorandum, "Subject: Discussion at the 425th Meeting of the NSC, November 25, 1959," November 25, 1959, *DDC*, 1992/366.

89. Taylor described in considerable detail the NSC discussions on basic national security policy that were held in 1958.

90. Samuel P. Huntington, "Strategy and the Political Process," *Foreign Affairs*, 38 (1960):285–99. This article drew the administration's attention. See Charles Haskins to Gray, memorandum, February 10, 1960, "Subject: Strategic Planning and the Political Processes," in Staff Memos (February–October 1960) folder, Special Assistant series, Subject subseries, WHO-OSANSA, DDEL.

91. Edwin L. Dale, Jr., "Eisenhower, in Budget Message, Stresses Need to Save Surplus, Asks Postal and 'Gas' Tax Rises," *New York Times*, January 19, 1960, p. 1.

92. Quoted in Jack Raymond, "Pentagon Chiefs Call U.S. Leader in Nuclear Arms," *New York Times*, January 20, 1960, p. 1. Gates's staff also gave him a list of responses to Taylor's charges. See attachment to memorandum, A. P. Toner to Goodpaster, January 16, 1960, Army (2)(3) folder, Subject series, Alpha subseries, WHO-OSS, DDEL.

93. Richard Russell (D-Ga.), the influential chairman of the Senate Armed Services Committee, stated: "From the way he [Gates] presented it, it was encouraging." Quoted in "Pentagon Chiefs," *New York Times*, CJCS Twining told President Eisenhower that "his own appearances and those of Mr. Gates had gone extremely well." Goodpaster, MCP [on January 25, 1960], January 26, 1960, *DDQC*, 1976/135A.

94. U.S. Congress, Senate Committee on Foreign Relations, *Executive Sessions of the Senate Foreign Relations Committee* (historical series), vol. 12 (1960), 86th Cong., 2d sess. (Washington, D.C.: GPO, 1982), p. 11.

95. Chairman William Fulbright (D.-Ark.) told Herter: "On the one hand, last year McElroy and Dulles and others told us that we had a pretty serious situation here. . . . If we can rest on Mr. Gates' assessment that the Russians are not as mean as we thought they were, and we should not be disturbed, that is going to make difficult a lot of our policies, and I think it is very disturbing to have what looks like quite a different approach." Ibid., p. 67.

96. In a subsequent confidential note to Senator Fulbright, Herter justified the change in the estimating process with respect to the size of Soviet missiles deployments: "Since last year, the intelligence community has received enough information about the Soviet missile program and about other factors which would affect a Soviet decision regarding the actual production and deployment of missiles to enable it to attempt an estimate of the probable program. This estimate, as Mr. Dulles has said publicly, does not represent a downgrading of the Soviet missile system from last year." Secretary Herter to Chairman, Senate Foreign Relations Committee, memorandum, in January 1960 folder, Nuclear History series, National Security Archive.

97. Gates also told the committee that this situation "will not produce a gap in our deterrent power" because of the range of nuclear weapons available to the United States. Quoted in John W. Finney, "U.S. Expects Soviet to Have 150 ICBM's in '61 for 3–1 Lead," *New York Times*, February 2, 1960, p. 1. For the background on this hearing, see Prados, *The Soviet Estimate*, p. 92.

98. McQuade to Nitze, memorandum.

99. See Prados, *The Soviet Estimate*, p. 94.

100. Goodpaster, MCP [on February 2, 1960], February 4, 1960, in Staff Notes—February 1960 (2) folder, DDE Diary series, ACW file, DDEL.

101. Elements of General Power's speech are remarkably similar to WSEG Staff Study No. 77; see Bottome, *The Missile Gap*, pp. 118–119. On the B-70, see Eisenhower telephone log, January 12, 1960, in Telephone Calls—January 1960 folder, DDE Diary series, ACW file, DDEL.

102. Gray, MCP [on Wednesday, February 3, 1960 at 8:45 A.M.], February 8, 1960, in June–December, 1959 (7) folder, Special Assistant series, Presidential subseries, WHO-OSANSA, DDEL; and Gray, MMP [on Wednesday, February 17, 1960, at 3:00, P.M.], February 19, 1960, *DDQC*, 1985/2146.

103. Quoted in Jack Raymond, "Juggling of Missile Data Is Charged by Symington," *New York Times*, January 28, 1960, p. 1. See also Prados *The Soviet Estimate*, pp. 90–95.

104. Felix Belair, Jr., "President Warns Rifts on Defense Imperil Morale," *New York Times*, February 12, 1960, pp. 1, 14. The president also attacked his critics at a political dinner on January 27: "In this kind of preachment, political morticians are exhibiting a breast-beating pessimism in the American system." Quoted in Bill Becker, "Eisenhower Calls U.S. 'Strongest'; Scores Pessimists," *New York Times*, January 28, 1960, p. 1.

105. Quoted in Jack Raymond, "Symington Says President Misled Nation on Arms," *New York Times*, February 20, 1960, p. 1.

106. Goodpaster, MCP [on March 11, 1960], March 15, 1960, in Staff Notes (March 1960) (2) folder, DDE Diary series, ACW file, DDEL.

107. Goldberg, *Strategy and Money*, p. 29; George Gallup, *The Gallup Poll: Public Opinion, 1959–1971*, vol. 3: *1959–1971* (New York: Random House, 1972), pp. 1639, 1661–1662; Kolodziej, *The Uncommon Defense and Congress*, pp. 320–321; and Huntington, "Strategy and the Political Process," pp. 286–288.

108. See "Army" and "B-70" briefing papers, in Briefing Papers, Air Pollution–Budget Policies folder, Position and Briefing Papers series, 1960 Campaign files, JFK Pre-presidential Papers, JFKL.

109. In *The Missile Gap* (pp. 137–145), Edgar Bottome shows how Kennedy broadened the missile gap issue into the overarching theme of U.S. power. See also Stephen E. Ambrose, *Nixon: The Education of a Politician* (New York: Simon & Schuster, 1987) chaps. 25–26.

110. "Republican Administration Errors and Miscalculations—Defense," Briefing Papers, Child Welfare–Depressed Areas folder, Position and Briefing Papers series, 1960 Campaign files, Pre-presidential Papers, JFKL. See also Desmond Ball, *Politics and Force Levels* (Berkeley: University of California Press, 1980), pp. 15–22.

111. Walter Barton Leach to White, letters, August 8 and October 25, 1960, Walter Barton Leach Papers, Harvard Law School Library.

112. Report, Committee on Defense Establishment to Kennedy, in Defense folder, Task Force Reports, Transition files, JFK Pre-presidential Papers, JFKL.

113. Report by PSAC Panel on Weapons Technology for Limited Warfare, August 3, 1960, "Weapons Technology for Limited Warfare," in Limited War, September 1959–December 1960 folder, WHO-OSST, DDEL. See also Gray, MCP [on Wednesday, August 24, 1960, at 9 A.M.], August 25, 1960, in Meetings with the President, 1960 2(7) folder, Special Assistant series, Presidential subseries, WHO-OSANSA, DDEL.

114. Lemnitzer to Gates, memorandum, December 9, 1960, "Subject: Deficiencies in the U.S. Posture for Limited Military Operations," *DDC* 1990/2584; and Douglas to Gray, memorandum, December 28, 1960, "Subject: Report on Possible Deficiencies in the U.S. Posture for Limited Military Operations," in Limited Military Operations, 1960 (3) folder, NSC series, Subject subseries, WHO-OSANSA, DDEL.

115. Rathjens to Kistiakowsky, memorandum, November 1, 1960, "Subject: Budget Issues in the Strategic Weapons Area," *DDQC*, 1984/1819. Skybolt is discussed more fully in the next chapter.

116. Memorandum, "Subject: Discussion at the 469th Meeting of the NSC, December 8, 1960," December 8, 1960, *DDC*, 1991/1995.

117. Report by the PSAC, "Review of the FY '62 Military Budget," *DDC*, 1987/2997.

118. Goodpaster, MCP [on December 5, 1960], December 8, 1960, *DDQC*, 1982/2881. See also National Security Council, "Missiles and Military Space Programs," NSC 6021, December 14, 1960, *DDC*, 1986/2855; "Revised Paragraph 5 of NSC 6021," *DDC*, 1986/2856; Douglas to Gray, memorandum, December 18, 1960, "Subject: Scope of Operational Capability of the Minuteman Program," *DDC*, 1986/1287; and Douglas to Gray, memorandum, December 27, 1960, "Subject: Scope of Operational Capability of the Polaris Program," *DDC*, 1989/83.

119. National Security Council, "Summary Evaluation of Our Actual and Potential Capabilities to Fulfill Current Commitments and Basic Objectives as Outlined in NSC 5906/1," NSC 6013, part 1, December 10, 1960, enclosure to memorandum, Robert Wade to Gray, December 14, 1960, in Meetings with the President, 1960 2(2) folder, Special Assistant series, Presidential subseries, WHO-OSANSA, DDEL.

120. Kennedy had probably told Eisenhower this on December 6 in a private meeting at the White House. Eisenhower's record of this meeting is included in Dwight D. Eisenhower, *Waging Peace* (Garden City, N.Y.: Doubleday, 1965), pp. 712–716.

121. Memorandum, "Subject: Discussion at 469th Meeting of the NSC." See also NSC Action no. 3-*c*, "Draft Record of Actions, 469th NSC Meeting, December 8, 1960," December 14, 1960, in Meetings with the President, 1960 2(2) folder, Special Assistant series, Presidential subseries, WHO-OSANSA, DDEL.

122. Farewell Radio and Television Address to the American People," January 17, 1961, *Public Papers of President Eisenhower, 1960–61* (Washington, D.C.: GPO, 1962), pp. 1039–1040.

123. George Quester makes a similar argument, although he asserts that the speech refers to all government spending, and not just the defense budget. See George Quester, "Was Eisenhower a Genius?" *International Security* 4 (1979):168.

124. Fred Greenstein, *The Hidden-Hand Presidency: Eisenhower as Leader* (New York: Basic Books, 1982), pp. 73–76.

125. John W. Sloan, *Eisenhower and the Management of Prosperity* (Lawrence: University Press of Kansas, 1991), p. 156.

126. McGeorge Bundy, *Danger and Survival* (New York: Random House, 1988), p. 349.

127. Greenstein, *The Hidden-Hand Presidency*, p. 57.

128. Bundy, *Danger and Survival*, pp. 334–349. In his excellent study of Eisenhower's management of the economy, John Sloan argues that Eisenhower became more

intransigent and less willing to work with Congress in his second term. "In assuming the role of a lonely moralist, he [Eisenhower] became a less effective politician and ended up achieving fewer of his objectives." See Sloan, *Eisenhower and the Management of Prosperity*, p. 161.

5. Strategic Force Planning during the Missile Gap Period

1. Robert Cutler to Eisenhower, memorandum, October 25, 1957, "Subject: SAC Concentration in U.S. and Reaction Time," in Security Resources Panel (2) (1957–1958) folder, NSC series, Briefing Notes subseries, WHO-OSANSA, DDEL.

2. Edmund Beard, *Developing the ICBM* (New York: Columbia University Press, 1976).

3. Security Resources Panel of ODM-SAC, *Deterrence and Survival in the Nuclear Age*, U.S. Cong., Joint Committee on Defense Production, 94th Cong., 2d sess. (Washington, D.C.: GPO, 1976), pp. 6, 23.

4. Goodpaster, MCP [on November 7, 1957], November 7, 1957, *DDQC*, 1979/331A.

5. Ignorance of the administration's examination of SAC vulnerability contributed to the panel's sense of failure after issuing its report. According to Goodpaster, "My impression—[the panel] came in expecting to give the big news & to startle—but had nothing that hasn't been on the President's mind & receiving his attention for weeks." Goodpaster, notes, November 7 (no year), in Military Planning, 1958–1961 (3) folder, Subject series, DOD subseries, WHO-OSS, DDEL.

6. "Strategic Air Command Bases and Aircraft," in Security Resources Panel (2) (1957–1958) folder, NSC series, Briefing Notes subseries, WHO-OSANSA, DDEL.

7. The ground alert program suffered from some weaknesses early in this period. Of the 157 bombers on ground alert in January 1958, only 18 had intercontinental range (12 B-52s and 6 B-36s). The remainder of the force, composed entirely of B-47s, was either located at vulnerable overseas bases or dependent on them for launching its attacks. This dependence of the ground alert force on overseas bases would continue to be part of SAC's plan, since 420 of the 515 aircraft (in mid-1959) would be B-47s. National Security Council, "Comments and Recommendations on Report to the President by the Security Resources Panel of the ODM," NSC 5724/1, December 16, 1957, *DDC*, 1986/2030.

8. *Alert Operations and the Strategic Air Command, 1957–1991* (Offutt Air Force Base: SAC Office of the Historian, 1991), p. 4.

9. Lieutenant General J. H. Atkinson to CINCSAC Power, memorandum, October 24, 1957, "Subject: SAC Alert Warning Times," Appendix CSUSAF to JCS, memorandum, January 16, 1958, "SAC Alert Warning Times," enclosure to note, Secretaries to JCS, January 21, 1958, JCS 1899/385, *DDQC*, 1981/155A.

10. The CIA later reduced the estimated range to several hundred miles. Power to White, letter, December 23, 1957, in 1957 Top Secret, General folder, White Papers, Library of Congress.

11. Ibid.

12. NSC 5724/1; and *Alert Operations and the Strategic Air Command, 1957–1991*, p. 4.

13. Report, "Specific Arms Control Proposals," enclosure C to "Initial Study of Arms Control Measures Affecting the Risk of Surprise Attack," WSEG Report no. 52, January 6, 1961 (WSEG-52), pp. 134–139, supplied in response to author's FOIA request.

14. CSUSAF to the JCS, memorandum, March 10, 1958, "Launching of the Strategic Air Command Alert Force," *DDQC*, 1981/157B. Several years earlier, Eisenhower had approved a directive which allowed for the automatic release of nuclear weapons from the AEC to the military in the event of a "Defense Emergency." There were three situations in which a United States commander could declare such an Emergency: a "major attack" by Sino-Soviet bloc countries on U.S. forces or on allied forces overseas; an attack on the United States; and the declaration of "Air Defense Readiness or Emergency." This constituted a major delegation of control over the nuclear stockpile, for it meant that by declaring a Defense Emergency, U.S. commanders could gain access to nuclear weapons without waiting for approval by the president or any other higher authority. See AEC-DOD, "Memorandum of Understanding for Transfer of Atomic Weapons," May 4, 1956, *DDC*, 1992/2401.

15. JSPC [Joint Strategic Plans Committee] to the JCS, report, May 1, 1958, "'Positive Control' Presentation to NSC," JCS 1899/402, in May 1958 folder, Nuclear History series, National Security Archive.

16. See ibid. Also memorandum, "Subject: Discussion at the 367th Meeting of the NSC, May 29, 1958," May 29, 1958, *DDC*, 1990/355; and "Supplemental Report on Items in Security Resources Panel Report," *DDQC*, 1985/37. For an extensive discussion of the dangers of airborne alert, see Scott D. Sagan, *The Limits of Safety: Organizations, Accidents, and Nuclear Weapons* (Princeton, N.J.: Princeton University Press, 1993), pp. 161–70.

17. During the Lebanon crisis in July 1958, the president placed SAC on alert. Within twenty hours after the order, SAC had almost 1,100 bombers prepared and on ground alert. SAC used information from this operation to develop plans for ground alert and dispersal programs. J. C. Hopkins and Sheldon Goldberg, *The Development of Strategic Air Command* (Offutt Air Force Base, Nebraska: SAC Office of the Historian, 1986); and Weapons Systems Evaluation Group, "First Annual Review of WSEG Report no. 23—The Relative Military Advantages of Missiles and Manned Aircraft (WSEG-23/1)," in Targeting and "The Relative Military Advantages of Missiles and Manned Aircraft" folder, NSC series, Briefing Notes subseries, WHO-OSANSA, DDEL, p. 14. See also CJCS Twining to McElroy, memorandum, August 21, 1958, "Subject: SAC Reaction Time," in DOD 3(1) folder, Subject series, DOD subseries, WHO-OSS, DDEL.

18. "Supplemental Report on Items in Security Resources Panel Report."

19. "Briefing for the President on SAC Operations with Sealed Pit Weapons," in DOD 2(9) folder, Subject series, DOD subseries, WHO-OSS, DDEL. Even though the Air Force claimed it had not been flying "war-ready" nuclear weapons, four aircraft had had accidents involving "operational nuclear weapons" in the preceding two years. Spurgeon Keeny to Killian, memorandum, January 30, 1958, "Subject: Accidents Involving Nuclear Weapons," *DDC*, 1991/2230. On sealed pit nuclear weapons, see Chuck Hansen, *U.S. Nuclear Weapons: The Secret History* (New York: Orion Books, 1988).

20. Goodpaster, MCP [on August 11, 1958], August 11, 1958, in JCS 1(4) folder, Subject series, DOD subseries, WHO-OSS, DDEL; and Twining to McElroy, memorandum, August 21, 1958.

21. Quarles to Eisenhower, letter, October 6, 1958, *DDQC*, 1981/286B.

22. Power to the JCS, memorandum, March 6, 1959, "Subject: Establishment of Airborne Alert," enclosure to JCS 1899/446, in CCS 3340 folder, JCS Records, RG 218, MMB, NA.

23. Two months earlier, Power told an Air Force Commanders Conference that airborne alert was needed because "the Communists believe that the U.S. once possessed the capacity to destroy Russia in an overwhelming surprise offensive but failed to do so; and that if and when the positions be reversed, the Communists would not make the same mistake." Major General Smart to Comptroller of the Air Force, memorandum, January 17, 1959, "Subject: 1959 Annual Commanders Conference," in Top Secret, General file, White Papers, Library of Congress.

24. General H. L. Hillyard to CINCSAC, memorandum, April 30, 1959, "Subject: Establishment of an Airborne Alert," SM 448-59, in CCS 3340 folder, JCS Records, RG 218, MMB, NA. The J-3 of the Joint Staff reported: "In general, CINCSAC states that the threat will commence by January 1960, whereas National Intelligence judges that the Soviets will achieve such a force on the order of a year later." See J-3 to JCS, "Report on the Establishment of Airborne Alert," April 23, 1959, enclosure B to JCS 1899/469, ibid.

25. Goodpaster, Memorandum for Record, April 7, 1959, in DOD 3(5) folder, Subject series, DOD subseries, WHO-OSS, DDEL.

26. Twining to Eisenhower, memorandum, August 13, 1959, "Subject: Dual Runways on SAC Bases," in CJCS 471.94 folder, CJCS Records, RG 218, MMB, NA.

27. Douglas to McElroy, memorandum, September 4, 1959, "Subject: Airborne Alert," in CCS 3340 folder, JCS Records, RG 218, MMB, NA.

28. Enclosure C to "Report on the Establishment of Airborne Alert," J-3 to JCS, October 20, 1959, JCS 1899/523, in CCS 3340 folder, JCS Records, RG 218, MMB, NA.

29. Burke to the JCS, Memorandum on Establishment of Airborne Alert, October 21, 1959, JCS 1899/525, in October 1959 folder, Nuclear History series, National Security Archive.

30. Burke apparently revised his memorandum to gain the support of Lemnitzer. Instead of advocating increased funding for invulnerable strategic weapons, the memorandum now blandly stated that "if funds in addition to those now allocated to the Defense budget are made available, they could be better spent in other ways." See "Views of the Chief of Staff, U.S. Army and Chief of Naval Operations on Establishment of Airborne Alert," appendix A to enclosure B to note by the Secretaries on Establishment of Airborne Alert, October 28, 1959, JCS 1899/527, in October 1959 folder, Nuclear History series, National Security Archive.

31. Joint Staff to JCS, Memorandum on the Establishment of Airborne Alert, January 26, 1960, enclosure to JCS 1899/543, in CCS 3340 folder, JCS Records, RG 218, NA; and Goodpaster, MCP [on March 18, 1960], March 26, 1960, in DOD 4(3) (March–April 1960) folder, Subject series, DOD subseries, WHO-OSS, DDEL.

32. White to York, letter, March 14, 1960, in 1959 Top Secret, General file, White Papers, Library of Congress.

33. Power to the JCS, letter, June 23, 1960, "Subject: Establishment of an Airborne Alert," in June 1960 folder, Nuclear History series, National Security Archive.

34. Secretaries to the JCS, note, August 25, 1960, "Airborne Alert Operations," JCS 1899/597, in CCS 3340 folder, JCS Records, RG 218, MMB, NA.

35. LeMay to Power, memorandum, July 19, 1960, "Recommended Actions for Increasing Capabilities and Readiness Posture of SAC Forces," in Top Secret file, 1960, White Papers, Library of Congress.

36. *Alert Operations and the Strategic Air Command, 1957–1991*, p. 7.

37. Such a discussion may have occurred in August 1958 just as Head Start I was about to begin. See Gray, Memorandum for Record, August 27, 1958, *DDC*, 1985/516; note that portions of the document have been deleted.

38. Enclosure C to WSEG-52, pp. 134–135.

39. See Robert McNamara to CJCS, memorandums of October 20, 1961, and November 17, 1961, "Subject: SAC Ground Alert Response." In CCS 3340 folder, JCS Records, RG 218, MMB, NA.

40. Jack Raymond, "Defect Is Found in B-47 Bombers; Modification Set," *New York Times*, May 3, 1958, p. 1.

41. Kenneth Patchin and James Eastman, *The B-52 Stratofortress*, vol. 3: *B-52 Deficiencies* (Tinker Air Force Base, Okla.: Office of the Historian, Oklahoma City Air Matériel Area, July 1961), pp. 86–87. Originally classified as secret, this monograph was declassified in February 1987 and is available at the AFHRC, MAFB. See also Marcelle Size Knack, *Post-World War II Bombers* (Washington, D.C.: Office of Air Force History, 1988), pp. 139–140.

42. Patchin and Eastman, *The B-52 Stratofortress*, 3:108–111.

43. Structural problems continued after 1961. In September 1962, 86 B-52G and B-52H aircraft were grounded because of structural problems in their wings. Power to LeMay, cable, September 4, 1962, in Messages and Cables (September 1962) folder, LeMay Papers, Library of Congress.

44. *Alert Operations and the Strategic Air Command, 1957–1991*, p. 79.

45. See Scott Sagan, "SIOP-62: The Nuclear War Plan Briefing to President Kennedy," *International Security* 12 (1987): 22–51.

46. Kenneth P. Werrell, *The Evolution of the Cruise Missile* (Maxwell Air Force Base, AL: Air University Press, 1985), p. 121. See also correspondence between LeMay and Lieutenant General Frank Everest, February 27, 1957, and April 12, 1957. In General Correspondence—Everest folder, LeMay Papers, Library of Congress.

47. Goodpaster, MCP [on March 20, 1958, at 3:00 P.M.], March 21, 1958, in Missiles and Satellites 2(3) folder, Subject series, DOD subseries, WHO-OSS, DDEL.

48. Memorandum, "Subject: Discussion at the 425th Meeting of the NSC, November 25, 1959," November 25, 1959, *DDC*, 1992/366.

49. Werrell, *The Evolution of the Cruise Missile*, p. 121.

50. Holaday to Twining, memorandum, January 9, 1959, "Subject: Advanced Air-to-Surface Missile," enclosure to note, Secretaries to JCS, January 15, 1959, JCS 2012/139, in CCS 4711 folder, JCS Records, RG 218, MMB, NA.

51. "Views of the Chief of Staff, U.S. Army and the Chief of Naval Operations on the Advanced Air-to-Surface Missile (AASM)," appendix A to memorandum, Twining to McElroy, April 17, 1959, "Subject: Advanced Air-to-Surface Missile (AASM)," JCSM-145-59, in CCS 4711 folder, JCS Records, RG 218, MMB, NA. A WSEG study had reported negatively on the technical difficulties involved in the missile's development. See memorandum, CSUSA, March 25, 1959, "Subject: Advanced Air-to-Surface Missile (AASM)," ibid.

52. "Views of the Chief of Staff, U.S. Air Force on the Advanced Air-to-Surface Missile (AASM)," appendix B to ibid.

53. Ibid.

54. White to DCS [Deputy Chief of Staff]/Plans and Programs, memorandum, September 3, 1959; and Major General Donnelly to LeMay, memorandum, September 9, 1959, "Additional Hound Dogs versus Less B-52's." In CS—Signed Memos folder, White Papers, Library of Congress.

55. Missiles Panel to PSAC, report, July 12, 1960, "Subject: The Skybolt Air-Launched Ballistic Missile Program," pp. 1–2, in Defense Program—FY 1961 Adjustments (1) folder, Subject series, DOD subseries, WHO-OSS, DDEL.

56. See Power to White, letter, September 2, 1959, in Command—SAC folder, White Papers, Library of Congress.

57. White to JCS, memorandum, November 24, 1959, "Advanced Air-to-Surface Missiles," JCS 2012/162, in CCS 4711 folder, JCS Records, RG 218, MMB, NA.

58. LeMay to White, memorandum, February 9, 1960, "Evaluation of the GAM-87A 'Sky Bolt'" [*sic*], in AF Council (January–June 1960) folder, White Papers, Library of Congress.

59. Rathjens to the Missiles Panel, memorandum, March 23, 1960, "Subject: May Meeting of the Missiles Panel," in March 1960 folder, Nuclear History series, National Security Archive.

60. Missiles Panel to Kistiakowsky, memorandum, May 16, 1960, "Subject: The Minuteman Program," *DDC*, 1988/529.

61. "Review of the FY '62 Military Budget," *DDC*, 1987/2997.

62. Goodpaster, MCP [on December 5, 1960], December 8, 1960. See also J. Holzapple to White, memorandum, December 9, 1960; and Roscoe Wilson to AFDDC, memorandum, December 19, 1960, "Dyna Soar and Skybolt Programs." In Missile/Space/Nuclear folder, White Papers, Library of Congress.

63. Kistiakowsky to David Beckler, memorandum, August 4, 1960, "Meeting with Secretary Douglas," in Missiles (6) folder, WHO-OSST, DDEL.

64. Kenneth Ciboski, "The Bureaucratic Connection: Explaining the Skybolt Decision," in John E. Endicott and Roy W. Stafford, Jr., eds., *American Defense Policy*, 4th ed. (Baltimore: Johns Hopkins University Press, 1977) p. 375; and Richard Neustadt to President Kennedy, report, November 15, 1962, "Skybolt and Nassau: American Policy-Making and Anglo-American Relations," Meetings and Memoranda series, Staff Memoranda subseries, National Security file, JFKL.

65. Missiles Panel to PSAC, report, July 12, 1960, "Subject: The Skybolt Air-Launched Ballistic Missile Program," pp. 1–2, in Defense Program—FY 1961 Adjustments (1) folder, Subject series, DOD subseries, WHO-OSS, DDEL.

66. Michael Brown, *Flying Blind* (Ithaca: Cornell University Press, 1992), pp. 210–211. For an excellent overview of B-70 development during the 1950s, see pp. 201–213.

67. Memorandum, "Chronology of Events Related to the Advanced Manned Strategic Aircraft Program, 1954–1965," p. 5, AFHRC, MAFB.

68. Ibid.; and Brown, *Flying Blind*, p. 217.

69. Memorandum, "Chronology of Events."

70. LeMay to White, memorandum, September 2, 1959, "Subject: Alternate Roles and Missions for the B-70 and F-108," Air Force Council–1959 folder, White Papers, Library of Congress; and Bernard Nalty, *The Quest for an Advanced Manned Strategic Bomber: USAF Plans and Policies, 1961–1966* (USAF Historical Division Liaison Office, August 1966), p. 2.

71. Power to White, letter, August 11, 1959, in August 1959 folder, Nuclear History series, National Security Archive. CSUSAF White even contemplated using the B-70 in the continuing conflict between the Navy and the Air Force. He explained that "the combination of B-52 or B-70 plus an air launched missile, preferably air launched ballistic missile, plus a reconnaissance satellite which picks up surface vessels, spells the end of surface Navy or Merchant Marine vessels in time of war." White to AFODC/AFDDC, memorandum, November 9, 1959, "Future Naval

or Merchant Vessel Survivability," in November 1959 folder, Nuclear History series, National Security Archive.

72. Goodpaster, MCP [on November 16, 1959], December 2, 1959, *DDQC*, 1980/107D. Two weeks earlier, Goodpaster had recorded: "The President said he is convinced that if we get into an all-out war both sides would attack the population centers of the other." Goodpaster, MCP [on November 3, 1959], November 6, 1959, *DDC*, 1988/1794.

73. Goodpaster, MCP [on November 18, 1959], January 20, 1960, *DDQC*, 1981/608. See also Robert Merriam, Memorandum for the Record, November 21, 1959, in Staff Notes—November 1959 (2) folder, DDE Diary series, ACW file, DDEL.

74. Goodpaster, MCP [on November 21, 1959, Augusta, Georgia], January 2, 1960, *DDQC*, 1980/108A.

75. Memorandum, "Chronology of Events."

76. Colonel Royal Allison to DCS [Deputy Chief of Staff]/P&P [Plans and Programs], memorandum, December 3, 1959, in December 1959 folder, Nuclear History series, National Security Archive.

77. Power to White, memorandum, January 11, 1960, "Subject: B-70 Flexibility," in January 1960 folder, Nuclear History series, National Security Archive.

78. Goodpaster, MCP [on January 25, 1960], January 26, 1960, in January 1960 folder, Nuclear History series, National Security Archive.

79. "The contractor estimates that the cost of the first hundred aircraft will be $4.1 billion, and experience with such estimates suggests that the actual cost may be nearly double, i.e. some $70 million per aircraft." Kistiakowsky to Eisenhower, memorandum, February 12, 1960, "Subject: Problems of the B-70 Project," in Kistiakowsky (2) folder, Administration series, ACW file, DDEL.

80. Memorandum, "Chronology of Events."

81. Quoted in Nalty, *The Quest for an Advanced Manned Strategic Bomber*, p. 4.

82. Brown, *Flying Blind*, p. 219.

83. Rathjens to Kistiakowsky, memorandum, November 1, 1960, "Subject: Budget Issues in the Strategic Weapons Area," *DDQC*, 1984/1819.

84. "Review of the FY '62 Military Budget."

85. Goodpaster, MCP [on December 5, 1960], December 8, 1960, *DDQC*, 1982/2881.

86. Nalty, *The Quest for an Advanced Manned Strategic Bomber*, p. 4.

87. Shortly after this budget request, the WSEG completed its massive report on strategic forces, in which it mildly stated that the B-70 "appears competitive" with Minuteman, although it downplayed the importance of post-strike reconnaissance missions, which the Air Force emphasized. Weapons Systems Evaluation Group, "Evaluation of Offensive Weapons Systems," Report no. 50, December 27, 1960, pp. 11–12, 16, in December 1960 folder, Nuclear History series, National Security Archive.

88. "NSC Actions Relating to the Aircraft Nuclear Propulsion Program," *DDC*, 1991/2734; Robert F. Little, *Nuclear Propulsion for Manned Aircraft: The End of the Program, 1959–1961* (USAF Historical Division Liaison Office, April 1963), pp. 1–10; and Brown, *Flying Blind*, pp. 194–196, 205.

89. Little, *Nuclear Propulsion for Manned Aircraft*, p. 11.

90. Goodpaster, MCP [on February 25, 1958], February 25, 1958, in February 1958 folder, Nuclear History series, National Security Archive.

91. John McCone and Quarles to Eisenhower, letter, January 2, 1959, *DDQC*, 1980/33A; and Little, *Nuclear Propulsion for Manned Aircraft*, p. 13.

92. Goodpaster, MCP [on January 8, 1959, at 8:30 A.M.], January 9, 1959, *DDQC*, 1976/218B.

93. Little, *Nuclear Propulsion for Manned Aircraft*, p. 17.

94. Ibid., p. 18.

95. By way of comparison, the takeoff weights of a B-52G and a B-1B are 488,000 pounds and 477,000 pounds, respectively.

96. Goodpaster, MCP [on June 23, 1959, at 11:40 A.M.], June 24, 1959, *DDQC*, 1976/218E.

97. Goodpaster, MCP [on October 8, 1957], October 9, 1957, in Missiles and Satellites 1(3) (September–December 1957) folder, Subject series, DOD subseries, WHO-OSS, DDEL.

98. Memorandum, "Subject: Discussion at the 339th Meeting of the NSC, October 10, 1957," October 11, 1957, in 339th NSC Meeting folder, NSC series, ACW files, DDEL; Cutler, memorandum, October 17, 1957, DDC, 1989/2982; and Goodpaster, MCP [on October 8, 1957], October 8, 1957, in DOD 2(2) folder, Subject series, DOD subseries, DDEL.

99. Gaither report, pp. 4, 6–7, 15.

100. Killian to Eisenhower, memorandum, December 28, 1957, *DDQC*, 1983/674.

101. In its analysis, the DOD cited the technical superiority of second-generation ICBMs as another reason to delay building a large force; NSC 5724/1, December 16, 1957, *DDC*, 1986/2030.

102. Memorandum, "Subject: Discussion at the 350th Meeting of the NSC, January 6, 1958," January 7, 1958, *DDC*, 1990/301. See also NSC, 5724/1; and "Comparison of Estimated U.S.-U.S.S.R. Missile Operational Capability," January 5, 1958, *DDC*, 1992/2587.

103. JSPC [Joint Strategic Plans Committee] to the JCS, "Report to the President by the Security Resources Panel of the ODM Science Advisory Committee (The Augmentation of the Present ICBM and Polaris/Submarine Weapons Systems)," March 26, 1958, JCS 2101/302, *DDQC*, 1981/158A.

104. Jacob Neufeld, *The Development of Ballistic Missiles in the United States Air Force, 1945–1960* (Washington, D.C.: Office of Air Force History, 1990), p. 187.

105. Gates to McElroy, memorandum, January 30, 1958, "Subject: Augmentation and Acceleration of the Fleet Ballistic Missile (Polaris) Program," appendix to note, Joint Secretariat to the JCS, February 6, 1958, JCS 1620/175, *DDQC*, 1983/174.

106. Kistiakowsky to Killian, memorandum, February 13, 1958, "Subject: Technical Progress and Actions Required in the Long Range Ballistic Missile Program," *DDC*, 1988/1140.

107. In advocating liquid missiles, Kistiakowsky pointed out: "If anti-missile-missiles become effective the large payload of Titan may become a necessity, to carry along sophisticated devices to overcome the defenses. . . . A retaliatory ICBM force made up of Titans with sophisticated nose cones and of solid propellant ICBM's with much lighter (and therefore not so sophisticated) nose cones, may prove to have an exceptional effectiveness." Ibid.

108. Goodpaster, MCP [on February 4, 1958], February 6, 1958, in Missiles and Satellites 2(1) (January 1958–February 1958) folder, Subject series, DOD subseries, WHO-OSS, DDEL.

109. Ballistic Missiles Panel to Killian, memorandum, March 4, 1958, "Subject: Whither Ballistic Missile Systems?" *DDC*, 1987/545.

110. Kistiakowsky to Killian, memorandum, February 28, 1958, in Missiles (1)

folder, WHO-OSST, DDEL; and Killian to Eisenhower, letter, March 8, 1958, *DDQC*, 1981/382A.

111. Goodpaster, MCP [on March 10, 1958], March 11, 1958, in Missiles (2) folder, WHO-OSST, DDEL.

112. Goodpaster, MCP [on March 20, 1958, at 3:00 P.M.], March 21, 1958, in Missiles and Satellites 2(3) (March 16, 1958–June 30, 1958) folder, Subject series, DOD subseries, WHO-OSS, DDEL.

113. Memorandum, "Subject: Discussion at the 363rd Meeting of the NSC, Thursday, April 24, 1958," April 25, 1958, *DDQC*, 1991/361.

114. "Program Increases Recommended by House in Procurement Areas in the 1959 Department of Defense Appropriation Bill," enclosure to memorandum, Stans to Eisenhower, June 18, 1958, in Program—Defense Budget FY 1960 (3) folder, Subject series, DOD subseries, WHO-OSS, DDEL.

115. Goodpaster, MCP [on June 23, 1958], June 23, 1958, in JCS 1(4) folder, Subject series, DOD subseries, WHO-OSS, DDEL.

116. McElroy to Eisenhower, letter, July 21, 1958; and Eisenhower to McElroy, letter, July 28, 1958. In Navy (4) folder, Subject series, Confidential file, DDEL.

117. Ballistic Missiles Panel to Killian, memorandum, July 18, 1958, "Subject: Status of Ballistic Missile Programs," *DDC*, 1988/1141.

118. Eisenhower told Twining: "[W]e must cut out some of the duplicatory weapons systems we have carried through the development stage." Goodpaster, MCP [on June 23, 1958], June 23, 1958.

119. Ballistic Missiles Panel to Killian, memorandum, July 18, 1958; and McElroy to Eisenhower, letter, December 3, 1958, *DDQC*, 1978/47C.

120. Robert Piland to Killian, memorandum, November 24, 1958, "Subject: Atlas-Titan," in Missiles (3) folder, WHO-OSST, DDEL; and Neufeld, *The Development of Ballistic Missiles*, pp. 187–189.

121. Ballistic Missiles Panel to Killian, memorandum, July 18, 1958; Goodpaster, MCP [on August 4, 1958], August 4, 1958, *DDQC*, 1975/155A; Piland to Killian, memorandum, November 24, 1958; and McElroy to Eisenhower, letter, December 3, 1958.

122. In the summer of 1958, the Air Force altered Minuteman to accommodate the state of solid fuel technology. As a result, the PSAC Missiles Panel informed Killian that "it is our opinion that this modified Minuteman weapon systems deserves a strong support to enable the development program to proceed as fast as the 'state of the art' allows. This done, an operational weapon system by '63–'64 seems quite probable. Such weapon system will truly be a 'second generation' advanced missile with increased reliability, resistance to an attack and of reduced installed unit cost." Ballistic Missiles Panel to Killian, memorandum, July 18, 1958.

123. Robert Piland to Killian, memorandum, November 24, 1958.

124. Kistiakowsky to Killian, memorandum, September 25, 1958, "Subject: Missile Problems," *DDC*, 1987/2319.

125. Neufeld, *The Development of Ballistic Missiles*, pp. 189–190.

126. According to a PSAC evaluation conducted in 1959, hardened targets would force the Soviets to reduce their missile CEP from 3.5 miles to 0.75 miles (a degree of accuracy unheard of in this era) if they wished to employ only one 1-megaton warhead with any confidence of eliminating the target. "Until such time as the Soviets have a highly advanced missile, one hardened aiming point added to our target system costs the Soviets from 10 to 20, or even more, missiles to negate. This factor applies to every missile of ours we can add to the system." Three Atlas

missiles hardened to 100psi would force the Soviets to allocate 200 rather than 5 ICBMs, to ensure the destruction of these targets. Brockway McMillan, "An Analysis of Technical Factors in the Strategic Posture of the United States—1956–64," March 4, 1959, in Air Defense (October 1958–August 1959) folder, WHO-OSST, DDEL.

127. Weapons System Evaluation Group, "First Annual Review of WSEG Report no. 23—The Relative Military Advantages of Missiles and Manned Aircraft," August 8, 1958, pp. 7, 12, 15, 18, in Targeting and "The Relative Military Advantages of Missiles and Manned Aircraft" folder, NSC series, Briefing Notes subseries, WHO-OSANSA, DDEL.

128. "Presentation by the Director, Weapons Systems Evaluation Group, to the National Security Council on the Subject of Offensive and Defensive Weapons Systems," October 13, 1958, supplied in response to author's FOIA request.

129. NSC Action no. 1994*b*, "Record of Actions by the NSC at Its 382th Meeting Held on October 13, 1958," *DDQC*, 1985/1817; and memorandum, "Subject: Discussion at the 382nd (Special) Meeting of the NSC, Monday, October 13, 1958," *DDC*, 1990/357; Gray, MMP [on October 16 at 11:00 A.M.], October 17, 1958, *DDC*, 1986/585; Gray to Twining, letter, October 30, 1958, *DDQC*, 1981/607A; and Killian to Gray, memorandum, October 29, 1958, supplied in response to author's FOIA request.

130. Enclosure C to report, J-5 to the JCS, "Minuteman Program," August 21, 1959, JCS 1620/270, *DDQC*, 1982/1567.

131. As discussed in Chapter 4, Symington pushed for the purchase of more Atlas missiles. Other senators sought an even faster acceleration for Minuteman. See letter of Swaetaer, February 11, 1959, *DDC*, 1987/544; and Mares to Killian, memorandum, February 12, 1959, "Subject: Missile Funding Delay," *DDC*, 1987/2082.

132. The other weapons on the list were Atlas, Titan, Polaris, the Nike-Zeus ABM, and reconnaissance satellites. NSC Action no. 2118, "Record of Actions by the NSC at Its 417th Meeting Held on August 18, 1959," *DDQC*, 1985/2693. The PSAC agreed to add Minuteman but thought "technical factors" prevented a more rapid acceleration. Ballistic Missiles Panel to Killian, memorandum, April 22, 1959, *DDC*, 1987/545.

133. Memorandum, "Subject: Discussion at the 406th meeting of the NSC, May 13, 1959," May 13, 1959, *DDC*, 1990/960.

134. NSC Action no. 2118; and Gray, MMP [on Friday, August 21, 1959, at 1:45 P.M.], August 24, 1959, in Meetings with the President, June–December 1959 (4) folder, Special Assistant series, Presidential subseries, WHO-OSANSA, DDEL.

135. LeMay to White, memorandums, May 12 and May 15, 1959, "Subject: USAF Objective Force Structure (1959–1970)," in Air Force Council folder, White papers, Library of Congress.

136. Ballistic Missiles Panel to Killian, memorandum, April 22, 1959, and Neufeld, *The Development of Ballistic Missiles*, p. 229.

137. J-5 to JCS, report, August 21, 1959, "Minuteman Program," JCS 1620/270, *DDQC*, 1982/1567. See also Burke to the JCS, memorandum, November 4, 1959, "Subject: JCS 1620/265—Minuteman Program," in CCS 4730 folder, JCS Records, RG 218, MMB, NA.

138. Major General Jacob Smart to Deputy Chiefs of Staff, memorandum, January 17, 1959, "Subject: 1959 Annual Commanders Conference," in Top Secret, General folder, White Papers, Library of Congress.

139. Robert Ginsburgh to White, memorandum, August 13, 1959, "Subject:

Minuteman Mobility," in August 1959 folder, Nuclear History series, National Security Archive. The planned railroad cars would have an apparatus that could raise the missile to launching position, allowing it to be fired directly from the train. Royal Allison to White, memorandum, October 23, 1959, in October 1959 folder, Nuclear History series, National Security Archive.

140. White to LeMay, memorandum, October 26, 1959; and Allison to AFXDC, memorandum, November 2, 1959. In CS Signed Memos folder, White Papers, Library of Congress.

141. CINCSAC Power informed White: "I feel that it is of utmost importance that the system be kept simple and inexpensive, so that it will be highly cost competitive with other systems. Along these lines, I have recently deleted a requirement for launch from any point in favor of the less expensive launch from pre-surveyed point system. Also deleted was the requirement for remote automatic retargeting for both the fixed and mobile Minuteman which will further reduce both complexity and cost." Power to White, letter, November 30, 1959, in November 1959 folder, Nuclear History series, National Security Archive.

142. Power informed White that "a force of three to five hundred mobile Minuteman is optimum for the useable trackage within the U.S." Ibid.

143. According to Goodpaster's record, McElroy told President Eisenhower that "he is resisting efforts to expand the program of the first generation of missiles. He thinks it is better to accelerate the Minuteman which can be railroad mobile." Goodpaster, MCP [on November 3, 1959], November 6, 1959, *DDC*, 1988/1794.

144. "Ballistic Missiles—A History," January 1960, in Missiles (5) folder, WHO-OSST, DDEL.

145. He added: "An example is that under the floor of the Command Post at the Vandenberg Air Force Base, controlling only three Atlas missiles, there are 30,000 relays." Kistiakowsky to Eisenhower, memorandum, August 4, 1959, *DDQC*, 1982/1291.

146. McElroy to James Douglas, memorandum, June 24, 1959, "Subject: Cost of ICBM Construction," in June 1959 folder, Nuclear History series, National Security Archive. According to Kistiakowsky, increasing base construction costs are "an unavoidable consequence of the requirement to have a 15 minute response time. Unfortunately, nothing much in the way of simplicity is gained by changing from 15 to, say, 20–25 minutes. To simplify base design drastically one needs to go to manual operations, and that probably means 1 to 2 hours response time." Kistiakowsky to Eisenhower, memorandum, August 4, 1959.

147. Jesse Mitchell, "Management Problems," May 14, 1959, in DOD (6) folder, WHO-OSST, DDEL.

148. President Eisenhower told Kistiakowsky that "it is possible that our philosophy of high reliability of missiles is not right. . . . it would perhaps be better to produce cheaper missiles and cheaper dispersed and hidden bases with a much lower level of reliability then [*sic*] we now demand." John Eisenhower, MCP [on August 4, 1959, at 11:00 A.M.], August 4, 1959, in August 1959 folder, Nuclear History series, National Security Archive.

149. Gray, MMP [on Friday, August 21, 1959, at 1:45 P.M.], August 24, 1959, in Meetings with the President, June–December 1959 (4) folder, Special Assistant series, Presidential subseries, WHO-OSANSA, DDEL.

150. A 1958 Air Force investigation found "that top level management did not have adequate knowledge or control of their operation." Schriever to White, letter,

"Titan ICBM Program," in Signed Memos file, White Papers, Library of Congress.

151. Schriever to White, letter, "Titan ICBM Program"; and Schriever to White, cable, December 1959, in Signed Memos file, White Papers, Library of Congress.

152. Ironically, the General Accounting Office had investigated the Air Force's management of ICBMs in late 1959 and early 1960. The GAO questioned the relationship between the Air Force and the Ramo-Woolridge/Space Technologies Laboratory. Although the company had provided invaluable assistance to the Air Force in missile development, the GAO criticized its "privileged" position. Assistant Director of the GAO Harold Rubin to Under Secretary of the Air Force Dudley Sharp, letter, November 5, 1959; and Comptroller General of the United Sates to Congress, "Report on Initial Phase of Review of Administrative Management of the Ballistic Missile Program of the Department of the Air Force (part 1)," November 1959. In DOD-AF-GAO (1960) (1) folder, WHO-OSST, DDEL.

153. Neufeld, *The Development of Ballistic Missiles*, pp. 190–194.

154. McElroy told Eisenhower that "we should keep on with Polaris at the rate of about three submarines per year." Goodpaster, MCP [on November 3, 1959], November 6, 1959.

155. See "Atlas, Titan, and Polaris Programs," January 6, 1960, *DDC*, 1991/3340.

156. When Navy officials briefed Budget Director Stans in mid-1959, they "informed [him] that the eventual objective of the Polaris program was 45 submarines, with 29 deployed at all times. With such a force, he was informed, we could destroy 232 targets, which was sufficient to destroy all of Russia. The total cost of such a program would be 7 to 8 billion dollars, and annual operating costs would be $350 million. An obvious question was suggested by this briefing—if Polaris could do this job, why did we need other IRBMs or ICBMs, SAC aircraft, and overseas bases? The answer he had received when he asked this question was that that was someone else's problem." Haskins, briefing note, February 10, 1960, *DDQC*, 1981/144D; and General Jacob Smart to Deputy Chiefs of Staff, memorandum, May 15, 1959, "USAF Tasks and Objective Force Structure." In May 1959 folder, Nuclear History series, National Security Archive.

157. Kistiakowsky to Eisenhower, memorandum, February 12, 1960, "Subject: Problems Involved in the Minuteman Program," *DDC*, 1987/2322. See also "Principal Technical Problems—Minuteman," February 12, 1960, in Missiles (1960) (5) folder, WHO-OSST, DDEL.

158. Memorandum, "Subject: Discussion at the 434th Meeting of the NSC, February 4, 1960," February 4, 1960, *DDC*, 1991/2038.

159. Douglas to Gray, memorandum, March 25, 1960, "Subject: Production of the Minuteman ICBM System and Related Operational Force Objectives," *DDQC*, 1983/834.

160. Memorandum, "Subject: Discussion at the 439th Meeting of the NSC, April 1, 1960," April 2, 1960, *DDC*, 1991/1971.

161. While "confused" about mobile Minuteman response time, the panel members concluded that one hour would elapse before a moving train could fire its missile. Further, "some reports we have seen indicate that after an hour the CEP of the system may be [deleted] and that in order to achieve the quoted [deleted] CEP, a further calibration period of about 2–3 hours may be required." Missiles Panel to Kistiakowsky, memorandum, May 16, 1960, "Subject: The Minuteman Program," *DDC*, 1988/529.

162. Long's presentation is available in the form of a memorandum to Eisenhower, May 3, 1960, *DDC*, 1987/528.

163. Goodpaster, MCP [on May 4, 1960], May 7, 1960, in May 1960 folder, Nuclear History series, National Security Archive.

164. Douglas to Eisenhower, letter, March 1, 1960, in Gates, 1959–1961 (4) folder, Administration series, ACW file, DDEL; and "Staff Notes no. 702," January 8, 1960, *DDQC*, 1985/1371.

165. "Cold launch" enabled missiles to be fired from submarines while they were beneath the surface without damaging the submarine. Goodpaster, MCP [on March 18, 1960], March 26, 1960, in DOD 4(3) folder, Subject series, DOD subseries, WHO-OSS, DDEL.

166. He favored this plan because it accommodated for technical adjustments after flight-testing and for the diversion of lead-time procurement items back to conventional submarines if the necessity arose.

167. Gates's concern about the existence of a missile gap probably arose from York's analysis of Soviet capabilities, which had been circulating for two weeks.

168. Goodpaster recorded that the president "commented that any emphasis on the particular date of 1963 seemed questionable to him, as a reversion to the game of guessing the 'year of greatest danger'." Goodpaster, MCP [on April 6, 1960], April 6, 1960, in April 1960 folder, Nuclear History series, National Security Archive.

169. See William Kaufmann to White, letter, February 18, 1960, in Missile/Space/Nuclear folder, White Papers, Library of Congress.

170. "Analysis of Cost-Effectiveness of Polaris and Minuteman," *DDQC*, 1981/404C.

171. Bureau of the Budget Military Division to the Director of the Bureau of the Budget, memorandum May 11, 1960, "Air Force Comparative Cost Analysis of Polaris and Minuteman"; and Bureau of the Budget Military Division to Stans, memorandum, May 11, 1960, "Subject: Cost comparisons, Polaris and Minuteman." In May 1960 folder, Nuclear History series, National Security Archive.

172. Eisenhower's determination to increase the size of the Polaris force is evidenced by his tepid response when informed of the Navy's latest force goal for Polaris: "[The president] had been told that the final objective of the Polaris program was forty submarines. Admiral Burke said the target was fifty submarines. The President said the number seemed to be going up. Admiral Burke replied that fifty had always been the largest number mentioned." Memorandum, "Subject: Discussion at the 453rd Meeting of the NSC, July 25, 1960," July 28, 1960, in 453rd NSC Meeting folder, NSC series, ACW file, DDEL.

173. Ibid.

174. Goodpaster, memorandum for Record, July 14, 1960, in Navy (5) folder, Subject series, Alpha subseries, WHO-OSS, DDEL; and Goodpaster, memorandum for Record, July 28, 1960, in DOD 4(5) folder, Subject series, DOD subseries, WHO-OSS, DDEL. Also Gates to Gray, letter, November 4, 1960, supplied in response to author's FOIA request.

175. George Rathjens of the PSAC staff wrote: "The Navy has repeatedly emphasized that they wish to go to 2,500NM to extend their operating range and not to increase the number of targets that they can cover; that is, they are clearly trying to make it appear that they do not intend to encroach on the SAC mission." Rathjens to Kistiakowsky, memorandum, November 1, 1960, "Subject: Budget Issues in the Strategic Weapons Area," *DDQC*, 1984/1819.

176. White to Lt. General Gerhart, memorandum, February 1, 1961, in Signed Memos (1961) folder, White Papers, Library of Congress.

177. Douglas to Chairman, Air Force BMC, memorandum, December 13, 1960, "Subject: Atlas, Titan, and Minuteman Fiscal and Development Programs," in Missile/Space/Nuclear folder, White Papers, Library of Congress. See also: National Security Council, "Missiles and Military Space Programs," NSC 6021, December 14, 1960, *DDC*, 1986/2855; "Revised Paragraph 5 of NSC 6021," *DDC*, 1986/2856; and Douglas to Gray, memorandum, December 18, 1960, "Subject: Scope of Operational Capability of the Minuteman Program," *DDC*, 1986/1287.

178. Weapons Systems Evaluation Group, "Evaluation of Strategic Offensive Weapons Systems," Report no. 50, December 27, 1960, December 27, 1960, p. 11, in December 1960 folder, Nuclear History series, National Security Archive.

179. This was done somewhat surreptitiously, Rathjens informed the PSAC, to "protect" both the WSEG and the PSAC. Rathjens to the Strategic Systems Panel, memorandum, September 23, 1960, *DDC*, 1988/1143.

180. George Rathjens, memorandum for record, September 21, 1960, "Subject: The General Balance of Strategic Forces and Their Relation to the FY 1962 Budget," *DDC*, 1988/561.

181. An exception is Herbert F. York, *Making Weapons, Talking Peace* (New York: Basic Books, 1987), p. 194.

182. Stephen Ambrose, *Eisenhower*, vol. 2: *The President* (New York: Simon & Schuster, 1984), p. 435.

6. Eisenhower and the Missile Gap in Perspective

1. On Kennedy's defense policies, see: Desmond Ball, *Politics and Force Levels* (Berkeley: University of California Press, 1980); John Lewis Gaddis, *Strategies of Containment: A Critical Appraisal of Postwar American National Security Policy* (New York: Oxford University Press, 1982); Lawrence Freedman, *The Evolution of Nuclear Strategy* (New York: St. Martin's Press, 1983); Desmond Ball, "The Development of the SIOP, 1960–1983," in Desmond Ball and Jeffrey Richelson, eds., *Strategic Nuclear Targeting* (Ithaca: Cornell University Press, 1986), pp. 62–70. The strategic bomber policies of the Eisenhower and Kennedy administrations are compared in Peter Roman, "Strategic Bombers over the Missile Horizon, 1957–1963," *Journal of Strategic Studies* 18 (1995):200–238.

2. A number of scholars on this subject generally agree with our assessment of Eisenhower's nuclear choices during the missile gap period. See McGeorge Bundy, *Danger and Survival* (New York: Random House, 1988), pp. 344, 349; David Alan Rosenberg, "The Origins of Overkill: Nuclear Weapons and American Strategy, 1945–1960," *International Security* 7 (1983):69; Herbert F. York, *Making Weapons, Talking Peace* (New York: Basic Books, 1987), pp. 193–196; Norman Graebner, "Conclusion: The Limits of Nuclear Strategy," in Graebner, ed., *The National Security: Its Theory and Practice, 1945–1960* (New York: Oxford University Press, 1986), pp. 275–304.

3. Walter LaFeber, *The American Age* (New York: W. W. Norton, 1989), p. 541.

4. Stephen E. Ambrose, *Eisenhower*, vol. 2: *The President* (New York: Simon & Schuster, 1984), p. 435. In his conclusion (p. 621), Ambrose criticizes Eisenhower for not making a greater effort to curtail the arms race.

5. Michael R. Beschloss, *Mayday: The U-2 Affair* (New York: Harper & Row, 1986), p. 230.

6. Ambrose, *Eisenhower*, 2:580.

7. Kenneth Thompson, "The Strengths and Weaknesses of Eisenhower's Leadership," in Richard Melanson and David Mayers, eds., *Reevaluating Eisenhower: American Foreign Policy in the Fifties* (Chicago: University of Illinois Press, 1987), p. 21.

8. Gregg Herken, *Counsels of War* (New York: Alfred A. Knopf, 1985), p. 116.

9. LaFeber, *The American Age*, p. 542.

10. Beschloss, *Mayday*, p. 367.

11. For a good summary of the subject of Eisenhower and arms control, see Charles Albert Appleby, "Eisenhower and Arms Control, 1953–1961: A Balance of Risks" (Ph.D. diss., Johns Hopkins University, 1987).

12. Theodore J. Lowi, "American Business, Public Policy, Case Studies, and Political Theory," *World Politics* 16 (1964):677–715; Paul C. Light, "Presidential Policy Making," in George C. Edwards III, John H. Kessel, and Bert A. Rockman, eds., *Researching the Presidency: Vital Questions, New Approaches* (Pittsburgh, Pa.: University of Pittsburgh Press, 1993); Light, *The President's Agenda: Domestic Policy Choice from Kennedy to Reagan (with Notes on George Bush)*, rev. ed. (Baltimore: Johns Hopkins University Press, 1991); and Steven A. Shull, ed., *The Two Presidencies: A Quarter Century Assessment* (Chicago: Nelson-Hall, 1991).

13. This is one of the central points made in Richard E. Neustadt, *Presidential Power and the Modern Presidents* (New York: Free Press, 1990).

14. Congress' role in foreign policy has changed somewhat in the post-Watergate period but it is clear that national security arguments can shield the executive from congressional review and oversight. On factors that have contributed to increased congressional activism, see James L. Sundquist, *The Decline and Resurgence of Congress* (Washington, D.C.: Brookings Institution, 1981).

15. Graham T. Allison, *Essence of Decision: Explaining the Cuban Missile Crisis* (Boston: Little, Brown, 1971), pp. 89–91, 171–175.

16. Robert J. Art, "Bureaucratic Politics and American Foreign Policy: A Critique," *Policy Sciences* 4 (1973):477–480.

17. Neustadt, *Presidential Power*, pp. 18–24.

18. The formulation of BNSP was usually based on an NIE prepared especially for that purpose.

19. Problems arose in the implementation of the subsidiary policy decisions that followed from the national strategy.

20. It would be possible for a president to suppress an intelligence report he disagreed with, but this could create political problems or undermine the estimating process over the long term by politicizing it.

21. For a discussion of formal and informal advisory structures, see Fred I. Greenstein, *The Hidden-Hand Presidency: Eisenhower as Leader* (New York: Basic Books, 1982), chap. 4; John Sloan, "The Management and Decision-Making Style of President Eisenhower," *Presidential Studies Quarterly* 20 (1990):295–314.

22. See James R. Killian, Jr., *Sputnik, Scientists, and Eisenhower* (Cambridge: MIT Press, 1982), especially p. 241; and Gregg Herken, *Cardinal Choices: Presidential Science Advising from the Atomic Bomb to SDI* (New York: Oxford University Press, 1992), chap. 7.

23. The dangers of presidential intervention in the policy process are examined in Larry Berman, *Planning a Tragedy* (New York: W. W. Norton, 1982).

24. John P. Burke, *The Institutional Presidency* (Baltimore: Johns Hopkins University Press, 1992); and Terry Moe, "Presidents, Institutions, and Theory," in Edwards, Kessel, and Rockman, *Researching the Presidency.*

25. Art, "Bureaucratic Politics and American Foreign Policy"; and Stephen Krasner, "Are Bureaucracies Important? (Or Allison Wonderland)" *Foreign Policy* 7 (1971):166.

26. Samuel P. Huntington, "Strategy and the Political Process," *Foreign Affairs* 38 (1960):297. This argument was made more recently by McGeorge Bundy in *Danger and Survival,* p. 349.

27. Norman Graebner, "Eisenhower's Popular Leadership," in Dean Albertson, ed., *Eisenhower As President* (New York: Hill and Wang, 1963), pp. 147–159; and Richard Neustadt, *Presidential Power,* chap. 8.

28. Neustadt, *Presidential Power,* pp. 52–55, 73–79, 135–141.

Index

Cornell Studies in Security Affairs

edited by Robert J. Art, Robert Jervis, *and* Stephen M. Walt

Political Institutions and Military Change: Lessons from Peripheral Wars, by Deborah D. Avant
Strategic Nuclear Targeting, edited by Desmond Ball and Jeffrey Richelson
Japan Prepares for Total War: The Search for Economic Security, 1919–1941, by Michael A. Barnhart
The German Nuclear Dilemma, by Jeffrey Boutwell
Flying Blind: The Politics of the U.S. Strategic Bomber Program, by Michael L. Brown
Citizens and Soldiers: The Dilemmas of Military Service, by Eliot A. Cohen
Great Power Politics and the Struggle over Austria, 1945–1955, by Audrey Kurth Cronin
Military Organizations, Complex Machines: Modernization in the U.S. Armed Services, by Chris C. Demchak
Nuclear Arguments: Understanding the Strategic Nuclear Arms and Arms Control Debate, edited by Lynn Eden and Steven E. Miller
Public Opinion and National Security in Western Europe, by Richard C. Eichenberg
Innovation and the Arms Race: How the United States and the Soviet Union Develop New Military Technologies, by Matthew Evangelista
Israel's Nuclear Dilemma, by Yair Evron
Guarding the Guardians: Civilian Control of Nuclear Weapons in the United States, by Peter Douglas Feaver
Men, Money, and Diplomacy: The Evolution of British Strategic Foreign Policy, 1919–1926, by John Robert Ferris
A Substitute for Victory: The Politics of Peacekeeping at the Korean Armistice Talks, by Rosemary Foot
The Wrong War: American Policy and the Dimensions of the Korean Conflict, 1950–1953, by Rosemary Foot
The Best Defense: Policy Alternatives for U.S. Nuclear Security from the 1950s to the 1990s, by David Goldfischer
House of Cards: Why Arms Control Must Fail, by Colin S. Gray
The Soviet Union and the Politics of Nuclear Weapons in Europe, 1969–1987, by Jonathan Haslam
The Soviet Union and the Failure of Collective Security, 1934–1938, by Jiri Hochman
The Warsaw Pact: Alliance in Transition? edited by David Holloway and Jane M. O. Sharp
The Illogic of American Nuclear Strategy, by Robert Jervis

The Meaning of the Nuclear Revolution, by Robert Jervis
The Vulnerability of Empire, by Charles A. Kupchan
Nuclear Crisis Management: A Dangerous Illusion, by Richard Ned Lebow
Cooperation under Fire: Anglo-German Restraint during World War II, by Jeffrey W. Legro
The Search for Security in Space, edited by Kenneth N. Luongo and W. Thomas Wander
The Nuclear Future, by Michael Mandelbaum
Conventional Deterrence, by John J. Mearsheimer
Liddell Hart and the Weight of History, by John J. Mearsheimer
Reputation and International Politics, by Jonathan Mercer
The Sacred Cause: Civil-Military Conflict over Soviet National Security, 1917–1992, by Thomas M. Nichols
Bombing to Win: Air Power and Coercion in War, by Robert A. Pape
Inadvertent Escalation: Conventional War and Nuclear Risks, by Barry R. Posen
The Sources of Military Doctrine: France, Britain, and Germany between the World Wars, by Barry R. Posen
Dilemmas of Appeasement: British Deterrence and Defense, 1934–1937, by Gaines Post, Jr.
Eisenhower and the Missile Gap, by Peter J. Roman
The Domestic Bases of Grand Strategy, edited by Richard Rosecrance and Arthur A. Stein
Winning the Next War: Innovation and the Modern Military, by Stephen Peter Rosen
Israel and Conventional Deterrence: Border Warfare from 1953 to 1970, by Jonathan Shimshoni
Fighting to a Finish: The Politics of War Termination in the United States and Japan, 1945, by Leon V. Sigal
The Ideology of the Offensive: Military Decision Making and the Disasters of 1914, by Jack Snyder
Myths of Empire: Domestic Politics and International Ambition, by Jack Snyder
The Militarization of Space: U.S. Policy, 1945–1984, by Paul B. Stares
The Nixon Administration and the Making of U.S. Nuclear Strategy, by Terry Terriff
Making the Alliance Work: The United States and Western Europe, by Gregory F. Treverton
The Origins of Alliances, by Stephen M. Walt
The Ultimate Enemy: British Intelligence and Nazi Germany, 1933–1939, by Wesley K. Wark
The Tet Offensive: Intelligence Failure in War, by James J. Wirtz
The Elusive Balance: Power and Perceptions during the Cold War, by William Curti Wohlforth
Deterrence and Strategic Culture: Chinese-American Confrontations, 1949-1958, by Shu Guang Zhang